California Forests and Woodlands

Figure 1. Sugar Pine (*Pinus lambertiana*), tallest and largest of
the world's pines, adds an elegant silhouette and long,
handsome cones to California's Midmountain
Forests.

California Forests and Woodlands

A Natural History

Verna R. Johnston

Photographs by the Author
Drawings by Carla J. Simmons

UNIVERSITY OF CALIFORNIA PRESS

Berkeley / Los Angeles / London

CALIFORNIA NATURAL HISTORY GUIDES
Arthur C. Smith, *General Editor*

Advisory Editorial Committee
Rolf Benseler
Ernest Callenbach
Raymond F. Dasman
Don MacNeill
Robert Ornduff

University of California Press
Berkeley and Los Angeles, California

University of California Press, Ltd.
London, England

Copyright © 1994 by The Regents of the University of California

Library of Congress Cataloging-in-Publication Data

Johnston, Verna R.
    California forests and woodlands : a natural history / Verna R.
Johnston ; photographs by the author ; maps and figures by Carla J.
Simmons
        p.   cm. — (California natural history guides ; 58)
    Includes bibliographical references (p.   ) and index.
    ISBN 0-520-08324-5 (cloth: alk. paper)
    ISBN 0-520-20248-1 (paper: alk. paper)
    1. Forest ecology—California.· 2. Conifers—California.
I. Title. II. Series
QH105.C2J59   1994
574.5'2642'09794—dc20   93-40070
                                                                  CIP

Printed in the United States of America

1 2 3 4 5 6 7 8 9

The paper used in this publication meets the minimum requirements
of American National Standard for Information Sciences—Permanence
of Paper for Printed Library Materials, ANSI Z39.48–1984. ⊗

This book is dedicated to warm memories of Amber Ellis, my special friend and field companion for more than thirty years.

Contents

Acknowledgments

I never guessed when I plunged into this project that it would take eight years to complete. Combining my own field experiences of more than four decades and those of other biologists with the latest in ecological research for this book has been a challenging commitment.

Many people helped along the way. I am grateful to the following colleagues who read individual chapters and made useful suggestions: John Sawyer, Jr., Humboldt State University; Bruce Bingham, PSW Forest and Range Experiment Station, Arcata; James Griffin, Hastings Natural History Reservation, Carmel Valley; Wayne Harrison, Calaveras Big Trees State Park; Robert Fry, Yosemite National Park; and Florence Lee Hawley, Bear Valley.

Information of value was made available by David Parsons, Sequoia/ Kings Canyon National parks; Michael Barbour, University of California, Davis; and Dolores Harley, Alex Horvath, and Gladys Shally, Calaveras Big Trees Association.

Mary Vocelka, formerly at the Yosemite Research Library, provided generous help in the early stages. Jean Beckner, Librarian, Honnold Collection, Claremont Colleges; Janet Mullen, Librarian, San Joaquin Delta College, Stockton; and the Stockton Public Library filled important informational gaps. The Library at the University of the Pacific, Stockton, offered a quiet mecca in which to think and research and write.

Field companions who added both zest and ideas to forest study trips included Nancy and Merle Akeson, Jean Beckner, Amber Ellis, Martha Estus, Alice and Addison Harris, Eleanor Hauselt, Florence Lee Hawley, Wally McGalliard, Wayne Roderick, Herb Scales, Gwen

Serriere, Winifred Tarpey, Ann and Anthony Trujillo, and Gordon Williams.

Essential support that kept me and the manuscript rolling came from other good friends: especially Vera and Edgar Brimberry, who provided an always-welcoming home away from home; and from Janet Mullen, Betty and Richard Prescott, Sharon Heitman, Norma and Eric Yeoman, Barbara Taylor, and Bill and Mary Scales.

My talented illustrator, Carla Simmons, was wonderful to work with throughout the project. Her drawings add an innovative artistic dimension.

Wally McGalliard took the author's portrait for the jacket. Dan Netz patiently initiated me into the world of word processing and stood by for the inevitable trouble-shooting. Hank Gregory, Jerry Chapman, Rick Gauuan, and Marie Sevier helped keep spirits high.

A very special acknowledgment must go to Arlene Mueller, who read and sharpened the entire manuscript, chapter by chapter. Her keen eye and succinct comments helped mold the original drafts into smoother, clearer prose. Her understanding of the ups and downs of writing eased many a rocky day.

Thoughtful editorial assistance was given generously by Arthur C. Smith, Ernest Callenbach, and Elizabeth Knoll.

To all of these, and to unnamed others, my deepest thanks.

The authorities consulted are listed under "Selected References" at the back of the book. Common and scientific names of plants are from *The Jepson Manual: Higher Plants of California* (1993), edited by James C. Hickman. In a few cases, where *The Jepson Manual* omits a common name or introduces a brand new one, an older established common name is included as an alternate.

This book follows the policy of the California Natural History Guides in capitalizing specific common names and specific plant communities for easier reading and comprehension. The scientific name is given the first time a plant or animal is mentioned; thereafter, only the common name is used. Both are indexed.

The book aims to bring hours of pleasurable, informative reading and an increased awareness of the priceless heritage of California's forests and their wildlife.

Verna Johnston
Camp Connell, California
1994

Introduction

From the misty Redwood Forests of its northwestern corner to the arid Pinyon Pine–Juniper Woodlands of its eastern high desert and the sunny Mixed Conifer Forests of the Sierra Nevada, California harbors some of the most beautiful and diversified coniferous forests on our planet.

Better known as pines, firs, cedars, cypresses, hemlocks, for example, conifers are trees that bear their seeds in cones and have needles or scalelike leaves. Very ancient as a group, they date back around 300 million years, long before the appearance of flowering plants.

Conifers dominate most California forests and woodlands today. More kinds and sizes of conifers grow here than in any other comparable region on earth. Of about fifty-six cone-bearing species and geographically distinct subspecies and varieties, three hold world records: the tallest—Redwood; the largest around—Giant Sequoia; the oldest—Bristlecone Pine.

The cone-bearing group includes twenty-one pines, eight firs, four junipers (large shrub or tree-size), three spruces, twelve cypresses, two hemlocks, two Douglas-Firs, one Redwood, one Giant Sequoia, and two "cedars." About half of these species are endemic to the state, occurring natively nowhere else. In numbers of species and the combinations in which they occur, as well as in magnificence of trees, California's coniferous forests stand unrivaled.

Oaks of about eighteen species, and many more varieties and hybrids, make up the other prominent wooded areas of the state. Oaks dominate many picturesque rolling woodlands of short, widely spaced trees in nearly pure stands. Some share ground with California Bay, Pacific Madrone, alder, maple and other hardwoods among the tall conifers. Some share dominance with the conifers.

The oaks in this book are covered primarily in their roles as part of the coniferous forest ecosystems. For details on oak woodlands, see Pavlik et al., *Oaks of California*, under "Selected References."

Forests and woodlands of all kinds encompass much more than trees. California's forests have evolved over time into intricate plant and animal communities interrelated in diverse, often amazing ways.

This book explores those interrelationships, revealing the roles played by wildlife, fungi, fire, flood, soil, weather, humans, and other elements in the ecology of the forests. It focuses, with admiration and depth, on the remarkable natural endowments of the conifers and targets the dangers they face in today's environment.

1

How to Recognize California's Conifers

Conifers, cone-bearing trees with needles or scalelike leaves, stay green throughout the year in California. Shedding needles, scales, or twigs a few at a time, they are never bare in winter. By contrast, many of their deciduous companion trees—maples, aspens, willows, dogwoods, alders, and some oaks—drop all their leaves in autumn. Conifers such as the larches and Dawn Redwood are deciduous but do not live in California forests.

Most conifers can be fairly easily identified by

a. where they grow,
b. the numbers and length of needles or patterns of scalelike leaves,
c. the shape and size of their cones,
d. sometimes, by their bark,
e. a combination of these traits.

But in each group there are some that are difficult to separate. Our look at California's coniferous forests will help you recognize the groups of conifers and most of the prominent individual species.

How to Tell a Pine from a Fir

Pines carry their needles in bundles, wrapped around at the base by a papery sheath. Some pines have two needles to a bunch,

Map 1. California topography from *California Mammals* by E. W.
Jameson, Jr., and Hans J. Peeters, illustrations by Hans J.
Peeters; University of California Press, 1988.

others three, still others four or five. A lone species, the Singleleaf Pinyon Pine, of eastern California, covers itself with stiff, slightly curved
single needles. Pines produce the familiar woody cones that hang down
from branches or adhere to stems, and which usually take about 2 years
to mature. Fallen cones last several years on the ground.

Firs, when young and perfectly symmetrical, have long been our traditional Christmas trees. Branches circle the trunk in whorls, generally
one whorl per year. Single needles line the branches and leave smooth
circular leaf scars when they fall. Fir cones sit upright near the ends of
upper branches, mature in one season, and fall apart on the tree unless
a squirrel nips them off before they are fully ripe.

Pines and firs of a number of different kinds occur together in California's forests.

Figure 2. Ponderosa Pine (*Pinus ponderosa*).

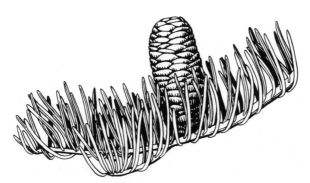

Figure 3. White Fir (*Abies concolor*).

Figure 4. Redwood (*Sequoia sempervirens*).

How to Tell a Redwood from a Giant Sequoia

California's two famous tallest and largest trees, the Red-wood and the Giant Sequoia, grow in such widely separate natural areas of the state that they need never be confused in the wild. The Redwood thrives in the coastal fog belt from Big Sur to southwestern Oregon. The forests which its tall reddish brown trunks dominate are unlike any others, moist dark groves much of the year, green with luxuriant ferns, the earth brown with fallen twigs that retain their sprays of flat needles and small 1-inch (3 cm) cones. A second kind of needle grows on some twig ends and higher branches—shorter, sharper, and awl-shaped.

The Giant Sequoia lives in scattered groves on the western slope of the Sierra Nevada, far across the Central Valley from its coastal cousin. The two species share only their reddish heartwood. The gigantic girth and brighter cinnamon bark of the Sierran tree set it apart as much as its scalelike leaves and its 2 to 3-inch (7 cm) grenadelike cones.

In the Giant Sequoia groves it is only Incense Cedar that you might at first confuse with the Giant Sequoias because both have cinnamon barks. But that uncertainty quickly sorts itself out as you walk the trails

Figure 5. Giant Sequoia (*Sequoiadendron giganteum*).

and absorb the massive presence of the Giant Sequoias, the spongy look of their bark, the spikiness of their leaves compared to the cedar's smooth, flat leaf sprays, vertically furrowed trunks, and lesser size.

How to Distinguish Cedars, Cypresses, and Junipers

These related members of the Cypress family all share scalelike leaves that overlap like shingles on a roof, but in very different arrangements—and they all bear small cones.

Incense Cedar, a major forest associate of pines and firs, grows widely throughout the main timber belts of the state, from the mountains of southern California to the Sierra Nevada and higher Coast Ranges. Its flattened leaf sprays, set in patterns of four encircling scales, separate it in northwestern California from the less common Western Red Cedar and Port Orford Cedar that share its habitat.

Figure 6. Incense Cedar (*Calocedrus decurrens*).

Figure 7. Monterey Cypress (*Cupressus macrocarpa*).

The cypresses native to California grow primarily in scattered "islands" of vegetation in the Coast Ranges, the Klamath Mountains, the Sierra Nevada, and southern California. Generally small to medium-sized trees with scalelike leaves that do not form flat sprays, they produce woody globe-shaped cones that usually open only when heated by direct sunlight or exposed to fire. Several species are quite rare. Best

Figure 8. Western Juniper (*Juniperus occidentalis*).

known are the windblown Monterey Cypresses of Point Lobos State Reserve.

The scalelike foliage and growth habits of junipers appear quite similar to those of cypress. But the junipers' purple or brown "berries," botanically considered cones, identify them readily. Geography usually settles the issue, as junipers and cypresses generally grow in very different places in California.

Separating Spruce, Hemlocks, and Douglas-Firs

In the moist northwestern coastal corner of California, Douglas-Fir and Redwood mingle with other tall giants in spruce-hemlock forests. Sitka Spruce stands out clearly with stiff, prickly, pointed needles, pungent when crushed, and papery cones 2 to 4 inches long (10 cm) that litter the ground. Western Hemlock bears flat needles that are neither prickly nor pungent and drops small 3/4-inch cones (2 cm).

The related Mountain Hemlock thrives in subalpine haunts of the Sierra Nevada and Klamath Mountains. It has shorter needles, soft to the touch, arranged starlike all around the branchlets. Its 2 to 3-inch

Figure 9. Western Hemlock (*Tsuga heterophylla*).

Figure 10. Douglas-Fir (*Pseudotsuga menziesii*).

cones (7 cm) often stand erect on their bases like pyramidal trees after falling. Mountaineers recognize the tree from far off by its drooping terminal shoot.

Both of the Douglas-Firs differ from the hemlocks in producing dangling cones with distinct bracts protruding beyond their cone scales.

The Bigcone Douglas-Fir in ravines of southern California mountain slopes has cones that grow 4 to 8 inches long (20 cm) with the bracts only mildly apparent.

The more widely distributed Douglas-Fir of the central and northern Coast Ranges and Sierra Nevada carries shorter 2 to 4-inch cones (10 cm) with very conspicuous three-forked bracts or "whiskers."

2

Redwood Forests

Majestically tall, impressively old, the Redwood (*Sequoia sempervirens*) forms forests of world-renowed beauty along the rainy Pacific Coast of northern and central California. From just south of Big Sur to slightly beyond the Oregon border, its tall spires dominate the lowlands and lower mountain slopes of the Coast Ranges in a roughly 10-mile-wide belt. Never continuous, the forests are intermittent in the south and relatively unbroken only from central Humboldt County north.

The largest, most magnificent groves thrive in the northwestern corner of California, many of them in state and national parks. Here the trees reach maximum heights in lush rain forest surroundings. Growing arrow-straight, their elegant reddish brown boles rise 200 to 300 feet high (90 m) from a forest floor often hidden by mosses, ferns, and herbs. Climbing up through the traceries of taller shrubs to form a green canopy far above, many individuals surpass 300 feet (90 m). A tree in Redwood National Park, known simply as Tallest Tree, holds the world's record at 368 feet (111 m). Its 10-foot (3 m) diameter is not unusual. Ten to 15-foot breadths (4.5 m) are common.

While Redwood trees are the tallest in the world and grow to venerable ages (1,000 to 2,200 years), they are not nearly as big around as their Sierra Nevada cousin, Giant Sequoia (*Sequoiadendron giganteum*), nor as old as the Western Bristlecone Pine (*Pinus longaeva*) of California's White Mountains. But Redwoods' old-growth forests emanate

12

Map 2. Redwood Forest.

very special moods unique to them. Somber and hushed in the fogs, cathedral-like when the sun radiates through canopy openings, they generate a serenity that slows the pace and takes you back in time.

Redwood family lineage does go back geologically to the Age of Dinosaurs. Fossils indicate that one hundred million years ago Redwoods of a dozen species spread widely over western North America, Europe, and Asia in a climate much milder than today's. Ice ages, volcanic eruptions, uplifts of mountain ranges, continental drift, and drastic climate changes all took their toll on population survival over the millennia.

By the end of the Ice Ages the Redwood species in North America had shrunk to two. The range of the two survivors dwindled to a fraction of its former size. Giant Sequoia continued to flourish only in isolated groves of California's Sierra Nevada. The Redwood, often

Figure 11. Redwood Forest. Redwood trees rise arrow-straight above the Winter Wrens that flit among the Redwood Sorrel and Western Sword Fern of the forest floor. Chestnut-backed Chickadees search tree leaves for insects. Salmonberry forms a shrub layer.

called Coast Redwood, was restricted to its narrow coastal belt. Today, by common usage, the name "Redwood" refers to the California coastal species alone.

Redwood Forests thrive in a mild maritime climate with heavy winter rains up to 100 inches (250 cm), frequent storms, and cool summers with dense night and morning fogs. Averse to salt spray, the trees generally live inland from the ocean just far enough to avoid it. Mendocino County has some exceptions. Following the wide river valleys as far as the ocean fogs penetrate, Redwood stands often occur up to 30 miles (50 km) from the coast in the north.

Wherever they grow, their massive crowns comb huge quantities of moisture from the mist and let it fall as "fog drip" to the forest floor, adding the equivalent of 12 inches (30 cm) or more of precipitation during the rainless summer months. Abundant water, high humidity, and the dense shade cast by the tall trees are the prime environmental ingredients with which Redwood Forests "create their own localized climate" (Becking 1982:1).

The trees themselves contribute heavily to the thick, life-sustaining litter of the forest floor. The litter's reddish brown hue comes largely from the branchlets of Redwood needles which fall to the ground in intact flat sprays when three or four years old. Mixed with these are the small, light brown cones, 3/4 of a inch long (2 cm), that mature in October and November of each year. The trees are prolific cone bearers, but the seeds that fall out through cracks in the cone scales usually show a very low germination rate. Seedling establishment may be poor unless floods or fires create the special conditions that improve seedlings' chances.

Such disturbances have occurred periodically over the centuries. Within recent decades severe floods hit Humboldt County in 1955, 1964, 1974, and 1986. The floods of 1955 and 1964 provided a vivid picture of what happens to Redwoods along river bottoms in northern California in extreme high water. Edward Stone and Richard Vasey documented the results.

The December 1955 and January 1956 floods swept sawmills, farms, and whole communities down the Eel, Klamath, and Van Duzen rivers, disrupting transportation and communication, burying buildings under thick mud. The old-growth groves that dominated the alluvial flats (terraces of soil) deposited by floods along the Eel River were greatly reduced in size.

"Trees were uprooted and carried downstream, and the herbaceous cover within the groves was buried under 4 feet (120 cm) of stream-carried silt." Bull Creek, a tributary of the Eel, "turned into a raging torrent, uprooted more than 300 (mature) Redwoods" and washed away some of the flats on which they had grown (Stone and Vasey 1968:157). Weakened stream banks caused later loss of 224 more large trees to erosion.

When Paul Zinke studied an exposed stream bank face along Bull Creek, using radiocarbon dating, he found that this sort of devastation was not new. "Fifteen major floods in the past 1,000 years have caused the deposition of sufficient silt (each forming a distinct profile) to raise the elevation of these alluvial flats more than 9 m [30 feet]" (Stone and Vasey 1968:158). Periodic erosion, followed by flooding and silt deposits on river flats downstream, has apparently been part of a natural Redwood environmental pattern for thousands of years.

Redwoods are well equipped to handle such disastrous environmental changes. When their trunks are buried beneath flood-carried silt, the trees sprout new roots that grow vertically upward from the buried ones, sometimes so strongly that the root tips come shooting through the surface. Later, a new more permanent horizontal root system spreads out from the trunk below ground and replaces the short-term vertical one. Since Redwoods may be buried many times in their long life of 600 to 2,000 years or so, they develop a multistoried root system, a new set for each flood they survive.

Redwood trees have a tap root, but it is their shallow roots, stretching wide and interweaving with those of neighboring trees, that give them a firm base. This adventitious system works well, but, in some areas, makes the trees vulnerable to windthrow unless flooding and siltation provide soil for new roots and added support.

The roots are sensitive, however, to compaction of the soil and may be suffocated by gravel and logging debris that does not allow soil aeration. Road building, clear-cutting, and skidding of logs in the past 100 years have drastically altered the natural pattern of flooding. Roads interfere with natural runoff on hillsides, causing landslides and heavy soil erosion. Clear-cutting, the removal of all, or nearly all, trees, adds to both erosion and landslides and makes the remaining trees more subject to windstorm damage. All of these increase the sediment in streams below.

When there is more sediment entering a river than the river can carry away, the excess piles up on flats and sandbanks in midstream. The main current is then forced toward the bank, undercutting trees growing

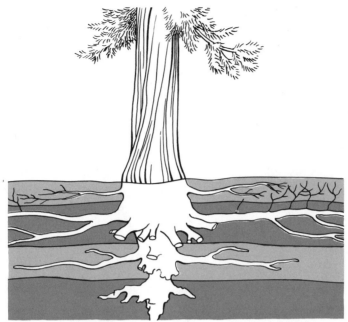

Figure 12. When floods bury Redwood trunks beneath silt, the trees sprout new roots closer to the surface. This tree has survived four different floods.

there. More than 100 old-growth Redwoods along the Avenue of the Giants, and ten in the Richardson Grove, were undercut and killed this way in the flood of 1986, when 23 inches (57 cm) of rain fell in 9 days along the South Fork of the Eel River.

The silt left by floods creates new Redwood habitat at the expense of that which is washed away. Siltation gives the trees the chance to grow more adventitious roots, which provide the support needed to grow taller, bigger, and older. And siltation provides the mineral seed-bed needed by seedlings.

In addition to floods, Redwoods have been exposed to fire over the centuries. And, here too, they are superbly adapted to survive. Their tough, fibrous bark, 6 to 12 inches thick (30 cm), does not burn easily. The large amount of water in their wood and their almost nonflammable pitch resist fire. But if a fire is severe enough, it will eat its way through the protective bark and open up the underlying wood to the next fire that comes along, eventually giving flames access to the heart of the tree. Most Redwood groves show some trunks with fire-blackened cavities.

Fire and flood both create the disturbances on the forest floor that encourage next spring's Redwood seedlings to get started. The exposed mineral soil, free from root-attacking fungi previously in the litter, gives them a better chance to survive if they are in enough sunlight to photosynthesize (produce food) and in enough soil moisture to keep their roots alive.

Redwood seedlings often become established on decaying down logs and upturned roots. Many die as the "punky" wood dries. Once established, however, even in extreme shade, a Redwood may survive in a suppressed condition for hundreds of years until an opening in the canopy permits it to shoot on up. Emanuel Fritz watched this happen to a 160-year-old Redwood that stood 100 feet tall in the understory beneath 300-foot-high giants. Once released by an opening far above, this tree, which had been growing at less than one percent a year, suddenly grew at 20 percent a year for the next 10 years.

Disturbances of any kind also bring out one of the Redwood's most remarkable traits, virtually unknown in conifers—its ability to sprout new stems and roots from burned, cut, or blown-down trunks. From the time that it is very young, a Redwood seedling carries latent buds around its base and along its trunk. When damaged or irritated, these dormant buds develop into leafy shoots and roots and, eventually, whole new trees. Masses of dormant buds, known as burls, bulge out like tumors on the trunks of some trees. Much sought after for their gnarled, twisted grains, burls produce beautiful designs and color effects when smoothed and polished.

The "fairy rings" of Redwood trees in many groves are the end result of prolific sprouting. An original monarch, badly burned, sprouts a dozen or more shoots all around its base. Three or four of these survive and grow into mature trees, circling the parent tree, which eventually dies. The survivors, in turn, if injured, produce sprouts around themselves, some of which survive. Over the centuries a ring of trees may stand around long-gone ancestors. This sprouting ability explains why there are still Redwoods growing on logged-over forests that have failed to regenerate new seedlings.

The natural floods and fires that have swept through the coastal forests over time have not only made seedling establishment possible but have actually helped the Redwoods maintain dominance over their competitors. Douglas-Fir (*Pseudotsuga menziesii*), Tan Oak (*Lithocarpus densiflorus*), Grand Fir (*Abies grandis*) and California Bay (*Umbellularia californica*), living on the surrounding slopes, all drop their seeds onto the same rich river-edge soils as the Redwood does. But all these are easily killed by fire, as they lack the protection of Redwood's thick bark.

The Tan Oaks of the Redwood Forest understory sprout vigorously after fires. They also tolerate shade and share with the Redwood a first-class defense against insects, fungi, and rot, namely wood and bark impregnated with tannin. But Tan Oak, along with Douglas-Fir, Grand Fir, and California Bay, succumbs to flooding and silting. Redwood survives both. Hence, on the river benches and floodplains of northern California, Redwood dominates the lowland areas. The tallest, oldest, most venerable Redwoods grow here.

But on the upland slopes, which are extensive, Redwood intermingles with the other conifers and hardwoods in forests that take on a true canopied look. Old-growth Redwoods and Douglas-Fir form the tall upper tier of the canopy. Douglas-Fir often dominates on shallow soils of steep slopes. Beneath their crowns grows an understory of Tan Oak, Pacific Madrone (*Arbutus menziesii*), and companion hardwoods. Below that tier a layer of shrubs and ferns proliferates—huckleberries, Western Sword Fern (*Polysticum munitum*), shrubby Tan Oak. Under these, on the forest floor, a ground cover of herbs flourishes—Redwood Sorrel (*Oxalis oregana*), Vanilla Leaf (*Achlys triphylla*), False Lily-of-the-Valley (*Maianthemum dilatatum*), and others.

Douglas-Fir is the most common associate of Redwood in the tall, upland forests—a handsome tree with slightly weeping branches and soft, blunt needles. Its fallen cones, with the identifying three-pronged bracts protruding between the scales like whiskers, give away its presence. Douglas-Fir accompanies Redwoods through their entire north-south range and becomes a major component of the drier forests to the east of the Redwood belt, as well as of the rain forests in Oregon and Washington.

The Grand Fir, an equally imposing tree, frequently grows to 180 feet high (54 m), with long branches and 1 to 2-inch needles (5 cm) that are a deep shiny green above and carry two bands of whitish stomata (breathing pores) on the lower surface. Grand Firs mix with Redwoods intermittently near the coast from Mendocino County northward but drop out a few miles inland.

The most abundant broad-leaved evergreen of the upland forests, Tan Oak, also known as Tanbark Oak, produces acorns like oaks but flowers and leathery leaves that resemble those of chestnuts and chinquapins. Indeed, this species is considered a possible connecting link between oaks and chestnuts. More than one hundred of Tan Oak's nearest relatives grow natively in Indomalaysia and Southeast Asia. Its bark once provided the chief commercial Western source of tannin. Its acorns still furnish food for Black Bears (*Ursus americanus*) and Black-tailed Deer (*Odocoileus hemionus columbianus*).

Yellow-rumped Warblers (*Dendroica coronata*), and other small birds, enjoy the black berries of another common understory tree, Wax Myrtle (*Myrica californica*). Nomadic Band-tailed Pigeons (*Columba fasciata*) sometimes swoop down to gorge on the orange-red berry crop of the reddish-barked madrones. Madrones, curving up to find scarce light among the taller trees, add white jug-shaped flowers in summer and large glossy leaves the year around to the forest scene. Tolerating moderate shade, they sprout readily from the stump after burning.

Another stump sprouter, Giant Chinquapin (*Chrysolepis chrysophylla*), occasionally becomes a fairly large tree in the understory of Redwood Forests. One of only two North American representatives from a primarily tropical Asian group, it bears narrow, leathery evergreen leaves up to 4 inches long (10 cm) which are dark green above and golden hairy below. Its distinctive burry fruit holds a sweet, purplish seed that is very popular with chipmunks.

Relatively uncommon needle-bearing trees that lack the seed cones of full-fledged conifers but grow among them include Pacific Yew (*Taxus brevifolia*) and California Nutmeg (*Torreya californica*). Yews form a part of the shady understory. Slender trees with thin bark that flakes off in rosy purple scales and downward sweeping branches of unequal length, they often present a somewhat scraggly profile. Their flat, two-ranked shiny green needles come to a point and slightly resemble those of Redwood. However, yew needles are pale green underneath, whereas Redwood's are whitish. In autumn the female trees produce seeds in juicy scarlet cups (arils) on the undersides of leafy twigs. The fruits appeal to many birds who eat the edible red flesh and scatter the poisonous seeds. The bark of yew yields Taxol, an important anti-cancer drug.

The attractive, dark green California Nutmeg grows singly or in scattered small stands, easily recognizable by the sharp, stiff prickle at the tip of each 1½-inch needle (4 cm). The crushed needles give off a peculiar odor, hence the name "stinking yew." The fleshy "fruit," borne on the separate female tree, looks a little like the commercial nutmeg of the tropics, though unrelated. It resembles a greenish plum with a large seed.

In Russian Gulch State Park on the Mendocino Coast, California Nutmeg attains possibly its largest size. A record tree near Fort Bragg stood 141 feet high (42 m) with a girth of over 14 feet (4 m) until cut by timber vandals in the early 1980s. Most nutmegs are much smaller. They stump sprout and grow almost as rapidly as Redwoods when young. It takes a special hunt to find them.

The search for an albino Redwood is much more difficult. These forest phantoms have been known since the 1860s, written about by Willis Linn Jepson and by Emanuel Fritz. Fritz, one of the world's authorities on Redwoods, called "Mr. Redwood," said that he had seen perhaps ten in his long life. Loggers, botanists, and others who have found them usually form an unofficial protection society by not revealing their sites.

The rare white Redwood is a true albino: its leaves and stems are totally white. Lacking the genes for chlorophyll synthesis, it survives by tapping the root system of its parent. It is usually one of many sprouts from a parent stump's "sucker patch." The largest one recorded stood 80 feet high (24 m) compared to the towering 200 to 300 feet (90 m) of green Redwoods. Most white Redwoods are much smaller, modest shrublike growths under 5 feet tall, though 20 to 30-foot individuals have been discovered. At any height, the unique albino is an unusual phenomenon in a Redwood Forest.

Redwood prefers sedimentary soil and will not usually invade other types. Where certain nonsedimentary soils intrude into a Redwood area, Coastal Prairies appear. In one such open grassy area in Prairie Creek Redwoods State Park, Roosevelt Elk (*Cervus elaphus roosevelti*) commonly graze.

Weighing well over 500 pounds, with massive antlers, the bulls are an impressive sight. During the fall rutting season, spectacular battles occur as the dominant bull clashes with rival males who challenge his possession of the harem. Once more widely ranging throughout northwestern California, elk now form fewer herds here. In coastal Oregon and Washington they still roam freely through dense forests and brush fields interspersed with meadows.

The greatest diversity of plant and animal life in the Redwood region occurs along the rivers and creeks flowing out to sea through breaks in the Coast Ranges. Here the sun penetrates along natural edges, and in the nesting and flowering months of May and June a whole riparian community bursts into life. Big-Leaf Maple (*Acer macrophyllum*) and gray-barked Red Alder (*Alnus rubra*) share the creekside habitat with berry thickets. Bogs mingle the spathes of Yellow Skunk Cabbage (*Lysichiton americanum*) with the veiled green of Giant Horsetails (*Equisetum telmateia*). Sometimes the horsetails reveal a yellow-green, 4-inch-long Banana Slug (*Ariolimax columbianus*) wrapped around a jointed stem, sleeping, breathing through a hole in its side, its eyes black dots on the ends of the larger tentacles.

Wilson's Warblers (*Wilsonia pusilla*) chant from the willows (*Salix* spp.). Dippers (*Cinclus mexicanus*) fly low over the water and disappear beneath the surface in their search for caddisfly and mayfly larvae. A Belted Kingfisher's (*Ceryle alcyon*) rattle downstream signals its hunt for unwary salmon or steelhead fingerlings.

The Winter Wrens (*Troglodytes troglodytes*) that were singing breathless songs from shaded stumps in late April are now busy feeding their fledglings. These demanding balls of dark brown feathers buzz and trill in and out of log tangles, landing on fern fronds that hardly ripple under their weight.

From an alder, Swainson's Thrush (*Catharus ustulatus*) notes spiral up the scale. Where the trail crosses the creek, half a dozen Chestnut-backed Chickadees (*Parus rufescens*) all bathe at one time in little puddles, descending from hillside Douglas-Firs and Tan Oaks with buzzy "dee-dee-dees" to splash in the watery depressions. An American Robin (*Turdus migratorius*) digs between creek pebbles, one eye on a chipmunk darting nervously in for a quick drink. And from up among the dark trees the haunting atonal notes of the Varied Thrush (*Ixoreus naevius*) echo through the forest.

In June the Red Elderberries (*Sambucus racemosa*) along the creek hang heavy with their annual crop of berries. The flat-topped white flower heads and huge leaves of Cow Parsnip (*Heracleum lanatum*) stand waist high alongside bushes of Cascara (*Rhamnus purshiana*) and Western Azalea (*Rhododendron occidentale*). Song Sparrows (*Melospiza melodia*) in the Thimbleberry (*Rubus parviflorus*) and California Blackberry (*Rubus ursinus*) bushes seem to be endlessly tuning up.

Along the creek banks and up the slopes, Man-root or wild cucumber (*Marah* spp.) twines among the shrubs, hiding well its emergence from a tuber as large as a man's body. Delicate saxifrages rim the wet spots. Red Columbines (*Aquilegia* spp.) mingle with clumps of Lady Ferns (*Athyrium filix-femina*), some ferns just unfurling the frond tip. others with crescent-shaped sori already developing on the under side.

The Five Finger Ferns (*Adiantum aleuticum*) carry anywhere up to sixteen "fingers."

The creeks and streams in the Redwood belt are themselves home to Silver Salmon (*Onchorhynchus kisutch*) and Steelhead Trout (*Onchorhynchus mykiss*) which cast shifting shadows on the gravelly sand and gray-bouldered bottom. During the rains of December, January, and February, the fish make their way from the ocean upstream to spawn. The adult salmon die after producing eggs and milt, but their young hatch in early spring and remain in the pools until next winter's rains raise the creek levels high enough for them to swim to sea. There they grow to mature size and in two years return to the streams in which they were born to continue the cycle.

Steelhead, ocean-going trout, do not die after spawning and may make three or four spawning runs in their lifetime. Their fingerlings spend one or two years in the creeks before heading out to sea.

These same creeks serve as important breeding arenas for many of the abundant amphibians in the Redwood belt. Pacific Giant Salamanders (*Dicamptodon ensatus*), Red-bellied Newts (*Taricha rivularis*), Tailed Frogs (*Ascaphus truei*) and others lay eggs in the cold waters, and their larvae grow to adulthood there.

Still other salamanders, as well as reptiles like the Rubber Boa (*Charina bottae*), the Ringneck Snake (*Diadophus punctatus*), which turns orange belly up when disturbed, and the Sharp-tailed Snake (*Contia tenuis*) which feeds entirely on slugs, live in rotting logs, beneath the bark of fallen trees, or burrow into the ground for the dry summer.

As with every ecosystem, a succession of flowers unfolds through the forest blooming season. The lovely Fairy Slipper or Calypso (*Calypso bulbosa*) ranks among the earliest. From February through May, its lone green basal leaf and short stalk topped by a delicate rose-pink orchid cover mossy logs in colonies or grow singly amid humus. It prefers the combined duff of Redwoods and Douglas-Fir but can survive in many mixed shade situations. And so low-growing is it that only a "belly view" will reveal its softly hairy throat and spotted lower lip.

The peak flowering period in the shaded Redwood Forests arrives in May. As the resident sword ferns shoot up fuzzy new fiddleheads, California Huckleberry shrubs (*Vaccinium ovatum*) produce pink and white flower bells beneath their shiny, saw-toothed leaves. Fragile-looking Sugar Scoop or Lace Flower (*Tiarella trifoliata*) covers acres with tiny white blossoms. Pink and white Star Flowers (*Trientalis lati-*

folia) rise on thin stalks above whorls of 3 to 6-parted leaves. Fairy Bells (*Disporum smithii*) hide creamy flowers under terminal parallel-veined leaves. Salal (*Gaultheria shallon*) overruns tangles with its thick evergreen foliage and rows of white urn-shaped flowers, and the California Rose-Bay's showy rose clusters brighten many a trail (*Rhododendron macrophyllum*).

Carpets of Redwood Sorrel spread heart-shaped leaflets everywhere over the spongy brown earth, leaflets so well adapted to the changing light and shade of the Redwood Forest that they droop dramatically in full sun, wilting from excessive water loss (transpiration). This very action reduces transpiration, allowing the leaves' vacuoles and intercellular spaces to fill up with water and expand again when back in the shade. When buried by silt, Redwood Sorrel grows vertical shoots to the new ground level, just as Redwoods do, and gradually recolonizes the forest floor. The plant is an indicator species of Redwood Forest.

The flowers of the forest floor could hardly be more distinct in size from the giants under which they grow. Yet, as Jepson has said, "the two, as living formations, belong together." The fossil record shows that they have come down through the ages, since before the Pleistocene epoch, linked for survival—the delicate flowers and the great red columns that provide their humus and shade (Jepson 1934:12).

Tiny ants share the link as well. Many of the understory herbs, such as Bleeding Heart (*Dicentra formosa*) and Western Trillium (*Trillium ovatum*), form seeds equipped with oil-bearing attachments that resemble dead insect parts. These attract scavenger ants, which carry off the seeds, feed on the oil attachments, and discard the seeds, dispersing the plants throughout the forest.

The shaded Redwood Forests house relatively few birds. An enormous quiet often settles among the trees, interrupted only occasionally by the croak of a Common Raven (*Corvus corax*), the tap of a woodpecker, the scold of a Winter Wren, the high, thin lisps of Golden-crowned Kinglets (*Regulus satrapa*), the harsh calls of Steller's Jays (*Cyanocitta stelleri*).

In recent years birders have added the loud, high "kree" notes of the Marbled Murrelet (*Brachyramphus marmoratus*) to the recognizable summer sounds of old-growth Redwood Forests. The home life of this small, chunky seabird was an ornithological mystery for decades. Most other members of its family—puffins, guillemots, murres, auks, and auklets—nest in colonies on steep sea cliffs or offshore islands. All feed

Figure 13. Marbled Murrelet, the unusual seabird that nests high
in old-growth Redwood trees and flies out to sea each
day to feed on fish. Its existence depends on the
ancient forests.

on ocean fish. But the Marbled Murrelet could never be found breeding
near any of its relatives.

Then, in 1975, a Marbled Murrelet nest was discovered on a mossy
limb 148 feet above the ground (44 m) in an old-growth stand of
Douglas-Fir and Redwood in Santa Cruz County. The nest contained
a single speckled green egg. It was in a tree 200 feet high (60 m),
located about 6 miles inland from the nearest salt water. Research since
then has pinpointed Marbled Murrelets' California nests primarily in
closed canopy, old-growth Redwood Forests. The birds' calls are heard
chiefly at sunrise and sunset as they fly over the tops of the trees or
circle them before flying below the canopy. Resembling flying cigars
with stubby wings beating rapidly, they vanish quickly into the tree
crowns. In 1992, the California Department of Fish and Game listed
the Marbled Murrelet as an endangered species in California. The bird's
future within the state depends on survival of its old-growth forest
habitat.

Most populations of birds and mammals of the Redwood Forests are found throughout the North Coastal Forests and Douglas-Fir/ Mixed-Evergreen Forests of the region as well. Nocturnal mammals, such as Raccoons (*Procyon lotor*) and Striped Skunks (*Mephitis mephitis*) raid the campgrounds and hunt along the creeks. Bobcats (*Lynx rufus*) stalk Dusky-footed Wood Rats (*Neotoma fuscipes*), chipmunks (*Tamias* spp.) and Brush Rabbits (*Sylvilagus bachmani*). The Douglas Squirrels (*Tamiasciurus douglasii*) that scamper so acrobatically through the trees during the day, feeding chiefly on conifer seeds, sometimes fall victims to the pounce of Martens (*Martes americana*). Actively hunting in early morning and late afternoon, these slender arboreal members of the weasel tribe, while rare, take their toll of Northern Flying Squirrels (*Glaucomys sabrinus*) discovered in tree holes, of birds and their eggs, and of any available prey.

Shrews (*Sorex* spp.) and Deer Mice (*Peromyscus maniculatus*) comb the forest litter. The high-strung shrews move about almost continuously, tracking insects, worms, and mice by scent. Shrews' tiny bodies, with a relatively large surface area and high metabolism, lose heat so rapidly that they must eat almost around the clock to stay warm and alive. Deer Mice, less frenetic feeders, consume a wide range of seeds, fruits, and insects.

In some of the shaded ravines inhabited by maples and alders, the unusual Shrew-Mole (*Neurotrichus gibbsii*) digs a network of runways deep in the forest litter. Mouse size, with the sensitive long nose of a shrew, it walks over the ground tapping it with the snout in its search for soil-dwelling pillbugs, centipedes, spiders, and insects. Its thick, heavy tail covered with stiff black hairs differs from the thin, mouselike tail of shrews.

All the interwoven plants and animals in the Redwood Forest ecosystem are tied intimately to the tall trees that shade them, the winter rains and summer fogs that water them, the mild maritime climate, the soil, the geography, and the accidents of history that brought them together in this singular natural home.

The Redwood's unique beauty, durability and usefulness as lumber has, however, proven to be its most vulnerable, almost fatal, feature. For during the last century, lumber companies have logged over 95 percent of the virgin Redwood Forests. Only about 2.5 percent remains in protected parks, and some of that, including the tallest trees of Redwood National Park, lies in lowland areas open to unprotected watersheds above.

Redwoods rank among the most resilient trees on earth. Adapted to naturally occurring floods and fires and able to resprout, they can turn a burned or cutover hillside into a green forest within a century. But during that century, the plants and animals dependent on the shade, soil moisture, shelter, and interrelated life of the old-growth ecosystem, will have died and vanished. And the second-growth forest replacement cannot begin to compare with the primeval beauty, the ambience, and the biodiversity of the 1,000-year-old forest gem long gone.

All of our remaining old-growth Redwoods, "standing tall" in their own special way among California's coniferous forests, deserve unequivocal protection from the bulldozer and the saw. We owe this one-of-a-kind Redwood Forest, and the future, no less.

3

North Coastal Forests

The forests north of Eureka get an increasing amount of rainfall, and Redwood trees share this ground with species that follow the Pacific Coast north to the temperate zone rain forests of Olympic National Park—Sitka Spruce (*Picea sitchensis*), Western Hemlock (*Tsuga heterophylla*), Western Red Cedar (*Thuja plicata*), and Douglas-Fir.

Remarkable conifers in every way, many spruce, hemlock, cedar and Douglas-Fir stand over 200 feet high (60 m), grow up to 15 feet in diameter (4.5 m), and live more than 500 years, occasionally 1,000 years in their prime habitats. The fogs, rain, overcast skies and moderate year-round temperatures produce damp, shady forests where the old trees space themselves out and form a canopy that beams soft filtered light on the forest floor far below. Much of the floor is a jungle of undergrowth. Mosses, lichens, and fungi cover boulders, mantle trees, and join the battle for space on fallen logs with ferns and young conifers. Deer Fern clumps (*Blechnum spicant*) take over the gentle slopes. Leather-Leaf Ferns (*Polypodium scouleri*) form minigardens in soil pockets on trunks. Red Huckleberry (*Vaccinium parvifolium*) and Salal grow out of stumps a dozen feet above the ground. Salmonberry (*Rubus spectabilis*), Blackberry, and Thimbleberry push up in scarce openings to form a shrub layer beneath the giant trees.

The scenic Loop Trail from the campground to the beach in Prairie Creek Redwood State Park traverses a fine sample of this coastal forest. Huge Sitka Spruce mingle with Western Hemlock and Grand Fir amid a rich understory. Hemlock is easily recognized by its drooping tips

Map 3. North Coastal Forest. Darker is primarily Spruce-Hemlock; lighter is intermixing Redwood and Douglas-Fir.

and short, blunt, variable needles with white bands below. Spruce needles are stiff and prickly.

Both hemlock and spruce produce tan papery cones—the hemlock's small, less than one inch (2 cm); the spruce's 2 to 4 inches (5 to 10 cm) long with slightly fluted scales. The barks look distinctly different—the spruce's with reddish brown scales that peel off; the hemlock's vertically furrowed. Grand Fir bears shiny, flat, dark green needles that set it apart from both.

The Lady Ferns and Wood Ferns (*Dryopteris expansa*) along the trail give no hint of the spectacular sight at trail's end, where Home Creek winds its way to the ocean through narrow, sheer-walled Fern Canyon, draped with thousands of delicately soft five-finger and polypody ferns (*Polypodium* spp.).

On adjacent Gold Bluffs Beach, the resident herd of Roosevelt Elk moves seasonally from Coastal Prairie to Coastal Strand to coniferous forest to Red Alder groves, in search of favorite foods. Nearby, Sitka Spruce groves show their predilection for closeness to the ocean. Along 2,000 miles (3,200 km) of Pacific shore, from northern California to Kodiak Island, spruce thrive within a few dozen (19 km) miles of salt water and follow coastal streams inland. Named for Sitka, Alaska, and also called Tideland Spruce, they seem completely at home in the cold, wet fogs near tidal waters.

On storm-battered cliffs and rocky islands off shore, Sitka Spruce take on the stout, craggy forms needed to survive windy blasts, but in the more protected deep forests, they frequently broaden into massive trees with great spreading limbs that reach nearly to the ground. A well-known specimen near Seaside, Oregon, measures 17 feet thick (5 m) and 216 feet tall (65 m) to a broken-off top. In the sheltered Hoh River valley of Olympic National Park, Washington, and in the Carmanah Creek watershed on Vancouver Island, many spruce surpass 300 feet (90 m). Fast growers, hence not always as ancient as their size might suggest, they commonly reach 400 to 700 years of age.

California's Sitka Spruce set no records on height or size, but they form some fine groves in the most southerly part of the tree's range, Mendocino County.

On the rugged promontories of Patrick's Point State Park, just inland from the bluffs where Bishop Pine (*Pinus muricata*) and Shore Pine (*Pinus contorta* ssp. *contorta*) compete for space, Sitka Spruce show their strong affinity for the ocean edge. Buffeted and pruned by the westerlies, they form both pure stands and mixed forests with Western Hemlock, Grand Fir, and Douglas-Fir.

The Spruce-Hemlock Nature Trail at Patrick's Point leads to some of the much touted "Octopus Trees" that form a regular part of the North Coastal Forests. When a tree falls in a spruce-hemlock forest, its prostrate trunk offers a sudden new available surface on which plants can grow—a bonanza to whatever can get there first. The competition in this moist environment is fierce, with mosses, liverworts, ferns, wildflowers, shrubs, and spruce and hemlock seedlings all in the running.

If the fallen log lies in a slightly open or disturbed area, spruce seeds dropping onto it and germinating in moist cracks or blown-in soil will quickly send roots down through the crevices, along the length of the log, over the sides, to the ground, even up into adjacent rotten stumps and then down, seeking moisture wherever they can find it. As the new tree grows above the log and around it, the roots firm up and thicken.

Over decades the log decays into a heap of soil on the forest floor, leaving the spruce standing in the clear above it on octopus-like "legs." And since Sitka Spruce are long-lived trees, they outlast many a "nurse log."

Hemlocks also favor these nurseries. In some spruce and hemlock forests more than 90 percent of the spruce and hemlock reproduction occurs on nurse logs. In dense shade such logs will be covered with hundreds of hemlock seedlings. Western Hemlock tolerates shade so heavy that very little else can grow in it. So even if spruce and Douglas-Fir rise high above in a mixed stand, if the shade is deep, it will be young hemlocks that carpet the rotting logs.

Hence, in most of the coastal forests that follow a natural succession, Western Hemlock may eventually emerge as the dominant tree. For even though it is slower growing and smaller in size than some of its associates, once it forms a dense canopy it cuts off light for nearly all except its own seedlings.

Sitka Spruce and Douglas-Fir regenerate well in disturbed areas or forest openings. Fire, flood, blowdowns, and logging all increase their chances for seedling establishment. And since such disturbances inevitably do occur, most North Coastal Forests are mixed stands. In many areas the mix includes Redwood.

Occasional companion trees include two elegant cedars, Western Red Cedar and Port Orford Cedar (*Cupressus lawsoniana,* formerly in the genus *Chamaecyparis*). Western Red Cedars reach into California locally as far south as southern Humboldt County. Their primary range stretches from the Alaska Panhandle south through western Washington and Oregon, and then inland into moist parts of the northern Rocky Mountains.

Striking trees when mature, they rise from heavily buttressed bases to heights of roughly 200 feet (60 m) on favored sites. Commonly growing 8 to 10 feet (3 m) in thickness, they may double that in Olympic National Park. The largest Western Red Cedars live close to 1,000 years. The trees retain their lower boughs, unless crowded, spreading aromatic branches that turn up slightly at the tip. They bear tiny scale-like leaves in flattened fanlike sprays, and, in season, 1/2-inch (12 mm) brownish cones. Caverns inside the buttresses of old Red Cedars furnish dry, fragrant dens for bears, skunks, Raccoons and wood rats.

Highly prized by the Northwest Coast Indians, who made fishing nets, sails, clothing and blankets out of the stringy, fibrous reddish brown bark, Red Cedars also helped out Lewis and Clark when their

Figure 14. North Coastal Forest. Sitka Spruce (left) rises on
octopus-like "legs" above the disintegrated log that
gave it birth. Its major associate, Western Hemlock,
stands at far right. Deer Fern and Vanilla Leaf cover
the forest floor. A Band-tailed Pigeon checks its
surroundings before eating the fruit of a Red
Elderberry bush. *Inset*: underground, knobby
mycorrhizae grow in and around conifer roots to
promote healthy tree growth.

expedition needed passage down the Clearwater River in northern Idaho to the Pacific Ocean. Their Nez Perce guides showed them how to make dugout canoes out of large Red Cedar trunks, hollowing out the trees first with fire, then finishing them off with stone tools. Nez Perce dugouts sometimes measured 60 feet long (18 m) and 8 feet wide (2.4 m) and could carry up to forty people. It's not surprising that the name "Canoe Cedar" came into common use.

Red Cedar's heartwood contains a natural fungicide that renders it highly resistant to decay. Shakes and fence posts hand-split from its wood last a century. For these and for its fragrant, naturally colored rough siding, Western Red Cedar has been much logged.

It is uncommon in California to find Western Red Cedar and Port Orford Cedar growing together, but in sections of Redwood National Park and Jedediah Smith State Park they join Redwood, Sitka Spruce, Douglas-Fir, and Western Hemlock in a diversified mix of North Coastal Forest.

Port Orford Cedar, a tall forest tree with fernlike leaf sprays on horizontally drooping branches, grows in some of the most varied habitats of western conifers. In addition to the fogbound North Coastal Forest, it thrives inland in open pine woods on harsh peridotite soils of Del Norte County, around streams and lakes in the Klamath region, and in mixed forests west of Dunsmuir.

An attractive tree, either when towering 175 feet (52 m) with most of the trunk clear of branches, or when softly foliaged to the ground in youth, it tolerates considerable shade. A slow grower, it may attain 600 years of age. Its light-colored, durable wood, with a spicy, ginger-like odor, has long been a favorite for cabinets, caskets, boats, and millwork. Consequently, its stands have been greatly reduced.

Port Orford's variable gene pool has yielded more than 200 cultivars over the years—weeping forms, silver- and gold-tipped varieties, some with bluish foliage, sprawling and columnar shapes—dozens of deviations, marketed by landscapers under the tree's other common name, Lawson Cypress.

The cedar's main foe, an introduced root-rotting fungus (*Phytophthora lateralis*), has in recent years decimated many Port Orford Cedars along the Oregon coast. Logging equipment seems to be the primary vector of the disease, transporting it from one cedar area to another. Regulations now require that equipment be cleaned prior to leaving a Port Orford Cedar site.

Where Western Red Cedar and Port Orford Cedar grow side by side,

Port Orford can usually be distinguished by the finer, dull blue-green sprays of foliage, white-flecked below, along with its thick, vertically furrowed reddish brown bark. Western Red Cedar's similar flat foliage tends to be shiny yellowish green or dark green, and its reddish brown bark thin and shreddy.

The streams and creeks that flow through the North Coastal Forests are often bordered by several kinds of broad-leaved flowering trees. The moss-draped trunks, branches and giant leaves of Big-leaf Maple overhang many a pool. In early May the new translucent leaves unfold with brilliant luminosity at about the same time that fragrant yellow 5-inch (12.5 cm) blossoms dangle from the boughs, enticing insects to pollinate them.

The fruit of Big-Leaf Maple, a double samara consisting of two nuts with 1 1/2-inch-long (4 cm) papery wings attached like a horseshoe, hangs on the trees through much of the winter—a treat for Douglas Squirrels, Evening Grosbeaks (*Coccothraustes vespertinus*), and other small birds and mammals.

The slender-limbed, airy Vine Maple (*Acer circinatum*) sets the understory aglow in autumn when its seven to eleven-lobed leaves turn a blazing orange-red. Black-tailed Deer and Elk delight in the tender new shoots of springtime.

Another stream dweller in the Redwood and North Coastal Forests from Santa Barbara to Alaska, the Red Alder, fills a special place in riparian ecology. The largest of four northwest species of alder, this handsome tree bears smooth gray bark marbled with light gray lichen and black bumps. Its leaves turn under slightly at the edges (revoluted). Tiny dark brown, conelike, female catkins hang from the branches year round, easily identifying it as an alder.

Its groves provide light, cheery oases amid the darker coniferous forests that surround them and also serve as natural firebreaks, since they are nowhere near as flammable as the conifers. Being deciduous, alders let in sun through their bare crowns much of the year, giving some understory plants time to get through their life cycles in spring before the trees' brown buds burst into the horizontal, open-leaf canopies of summer.

Ecologically, Red Alders rate as gold mines. Wherever they live, they fertilize the soil. The thousands of nodules that cling to their roots contain nitrogen-fixing bacteria able to convert atmospheric nitrogen, useless to plants, into organic nitrogen compounds which plants can use. These compounds travel up the alder stems to the leaves as amino acids. When the leaves fall to the forest floor each autumn and decay, they add usable nitrogen to the soil.

Alders enrich forest soils just as alfalfa and other legumes fertilize farmlands, and they do it naturally, without the pollution problems of water-soluble chemical fertilizers that wash into streams and promote algal growth (eutrophication).

Phenomenally fast-growing trees, alders play a prominent role in the forest succession. Alaskan alders are well known for their pioneering on the raw infertile soil left by retreating glaciers. Within 70 years after an Alaskan glacier has melted back on its borders and exposed dirt beneath, an alder forest may move onto the newly released land. The alders stabilize the sterile soil with a shallow root network, add around a ton of nitrogen per acre to its upper 18 inches (45 cm), and put down a rich layer of humus and leaf mold.

As the alders gradually make the land livable, Sitka Spruce seeds invade it from adjacent forests. As the spruces grow, the alders die out, intolerant of the newcomers' shade. By 170 years after the retreat of the glacier and the first exposure of land under it, a Sitka Spruce forest with trunks over 100 feet high (30 m) will fill the Alaskan area.

Something similar happens in logged or burned areas of the wet northern California coast. After an initial bushy growth of sword ferns, Salmonberry, Thimbleberry, Salal and various herbs, Red Alders come in, spread by their prolific, lightweight, windblown seeds. Quickly outgrowing the competition, they shoot up more than 3 feet (.9 m) a year for the first 10 years.

As time passes and the alder's nitrogen and litterfall enrich the ground, the conifers that have invaded the alders begin to outpace them and, by the sixtieth year or so, the conifers attain dominance over their short-lived rivals.

Coniferous forests are not noted for an abundance of food, but the birds and mammals living in these North Coastal Forests find a variety of foods available if they like the parts of conifers—needles, buds, twigs, bark, seeds, cones. The seeds of spruce, hemlock, Grand Fir, and Douglas-Fir, though small, all contain essential fats and proteins. Douglas Squirrels chop off thousands of seed-laden cones each autumn. Those that they don't cache alongside logs or in other hiding places go to hungry chipmunks, Deer Mice, and wood rats.

Birds such as Red Crossbills (*Loxia curvirostra*) and Pine Siskins

(*Carduelis tristis*) settle in flocks on conifer crowns when the cones are ripe. Crossbills' crossed beaks work with precision in picking seeds out of cones, and they join the Siskins in banqueting on fresh young needles, buds, and insects in the treetops.

Resinous conifer needles, when young and tender, go down many wild palates. Black-tailed Deer and Roosevelt Elk like them in the spring, along with new shoots of Sitka Spruce and other young conifers, and Blue Grouse (*Dendragapus obscurus*) count on them as a winter staple. Red Tree Voles (*Arborimus longicaudus*) can spend their entire lives in Douglas-Fir, feeding on nothing but its needles.

Conifer bark has its adherents; it provides a reliable food source in winter for Porcupines (*Erethizon dorsatum*) and Mountain Beavers (*Aplodontia rufa*), and for deer and squirrels if nothing else is available.

The rarely seen Mountain Beavers, short, heavy-bodied, dark brown, almost tailless rodents resembling muskrats, come out to feed chiefly at night. Poorly named, they are neither beavers nor entirely mountainous, though some live in the montane forests of the Sierra Nevada, as well as in the Mono Basin.

Primitive mammals, the only one of their kind left, they occur nowhere else in the world except in the Pacific Coast states. Preferring moist forests near streams, they dig labyrinthine burrows under dense tangles of ferns and berries and cut tunnels through thickets. Here, in a virtually impenetrable jungle, they live a vegetarian's life, feeding on practically all of the leafy plants available, climbing alders and other small trees to prune twigs for their caches, drying them outside the tunnels.

Unlike gophers, they leave no dirt piles at their entranceways, but in extremely wet areas, may shield the entrance with sticks, leaves, and fern fronds. The tangled undergrowth protects them somewhat from predators such as Mountain Lions or Cougars (*Felis concolor*), Bobcats, skunks, and Coyotes (*Canis latrans*), but not from weasels (*Mustela* spp.) and Mink (*Mustela vison*), which can follow them into the most intricate network of subterranean passages.

Mountain Beavers frequently share their tunnels part of the year with the coastal forests' largest salamander, the Pacific Giant Salamander. Nearly a foot long, half of it a vertically flattened tail, with a black-blotched shiny gray skin and eyes that protrude like bronzy marbles, the Pacific Giant Salamander lives as secretively as the Mountain Beaver, catching insects and even small mice with lunges that imbed its sharp teeth firmly in its prey.

Figure 15. The elusive Mountain Beaver digs burrows through
tangles of ferns and berry bushes, emerging chiefly at
night to feed on plants.

During the courtship season, the Giant Salamanders move to nearby
streams to pair off and breed. The male deposits his sperm in gelatinous
capsules (spermatophores) on the stream bed. The female picks up the
capsules with her cloacal lips and brings them into her body where they
fertilize her eggs internally.

After she lays a clutch of up to 200 fertilized eggs in a cavity in the
stream bed, she guards them till they hatch in about six months. Once
the gilled larvae take off on their own, she returns to her tunnels in the
moist forest floor. Here she may well encounter voracious shrews, along
with chipmunk and Deer Mice competitors, all avid for the beetles,
earthworms, termites, carpenter ants, pillbugs, millipedes, springtails,
and other small fry of the ground world.

Robins, Varied Thrushes, Winter Wrens, Swainson's Thrushes, Steller's Jays, and other birds also hunt the forest floor. Some of them find
juicier pickings in the berries that ripen from summer to fall. Salmon-
berries come early, producing what looks like salmon-colored black-
berries in May and June when a few of the deep rose-pink flowers still
nod on the tall prickly-stemmed bushes. Thimbleberries' raspberry-like

fruits, tasty but mild, develop from large attractive white flowers on medium-height, unarmed shrubs with leaves that feel like velvet.

Currants and gooseberries of several species (*Ribes* spp.) produce berries devoured by many birds and mammals. The smooth stems of currants allow easier handling than the spiny stems and fruit of gooseberries, but chipmunks and other wildlife manage to avoid the bristles and salvage the juicy contents.

Black Bears often gorge themselves on red and black huckleberries, important fattening foods, and also favor the dark purple berries of Salal, one of the most common cover shrubs in the region. A host of less obvious foods grow in forest openings or in the forest interior for patient seekers.

The early rains of autumn bring a new source of food suddenly into prominence on the forest floor. Mushrooms push up through the needle-strewn humus everywhere—Amanitas, Boletes, Laccarias, Rhizopogons, Armillarias—dozens of species in abounding variety and numbers.

Deer gorge on them, seemingly unaffected by the toxins in Amanitas and others that are deadly to humans. Chipmunks nibble the mushroom caps and carry away cheekfuls to their nests. Deer Mice, the most abundant small mammals in the woods, add mushrooms to their wide-ranging menu. Douglas Squirrels wedge chunks beneath loose bark or in forks of branches to dry before carrying them off to a cache.

And while all this activity is going on above ground, totally different events are happening below. For the mushroom itself merely constitutes the fruiting body of a fungus whose business end is a vast network of threads beneath the surface. These threads (hyphae) reach out into the soil in all directions, producing chemicals that selectively digest and absorb dead organic matter and turn it into energy.

Some fungi are parasitic, but most are not. Since fungi lack the chlorophyll of green plants, they cannot photosynthesize their own food and must depend on energy from their surroundings. Eventually some of the energy generated underground by their hyphae thrusts a mushroom up through the surface and produces spores on the underside of its cap. When the spores ripen and either drop or are wind-blown or animal-carried to favorable places, they sprout hyphal threads and begin a new plant.

While fungi literally "eat" their way through their environment underground, many of them do not limit their domain to what their own hyphae can provide. Plant physiologists have known for over a century that some fungi maintain intimate relationships with certain trees. Recent research has made the details more clear.

These particular fungi wrap their threads in a mantle around the host tree's tiny growing root tips or penetrate the rootlets in a mutually beneficial (symbiotic) relationship known as a mycorrhiza (my-co-rise-a) or fungus root.

The fungal partner of this symbiosis acts as a highly efficient extender of the host tree's root system, absorbing minerals and water from the soil in areas the roots would not ordinarily reach and transferring them to the tree. The fungal partner stimulates the host to produce new, more elongated root tips of increased life span. It brings in to the tree nutrients such as phosphorus, essential for growth and development but often present in amounts too low for the tree's own roots to absorb. The same happens with copper and zinc. The fungal partner improves water intake. Some fungi even give off compounds that protect their host tree from diseases. Other detoxify soil substances.

In return for all these benefits, the tree gives the fungal partner simple sugars made in its green leaves—food energy to continue the arrangement.

Conifers, oaks, cottonwoods, willows and alders all have fungal partners on their roots. Some fungi are highly host-specific in their tree partners; others are more wide-ranging. Over one hundred species of fungi form partnerships with Douglas-Fir. The beautiful, deadly, red-and-white spotted *Amanita muscaria* mushrooms are partners with all the conifers in North Coastal Forests and with most California pines, whereas the fungi specific to Red Alder will not team up with any conifer.

All the forms of mycorrhizae seem to do the same thing for the partners involved—promote healthy growth. Interrelationships such as these obviously have evolved over long periods of time.

The old rhyme "There's a fungus among us" was never more true than in North Coastal Forests. Dig anywhere and masses of moldlike fuzz, fine roots with knobby stubs on them, and strands ranging from white to lemon to tan to orange to pinkish, and even jet black, reveal mycorrhizae at work. All in the top soil layer, they keep the forest healthy.

Many of them interlace, connecting different trees. James Trappe once traced a single hypha emerging from a Douglas-Fir fungus root connection in a rotten log. The hypha reached out for more than 6 feet

(2 m) and had over 120 lateral branches or fusions with other hyphae. Thirty-four of these branches connected to other mycorrhizae on the same tree, nine to mycorrhizae of hemlock roots growing in the same log.

Many plants cannot survive at all without their fungal connection. When Monterey Pine (*Pinus radiata*) seeds from California were introduced into nurseries in Western Australia in 1927, they failed miserably until soil was added from areas where the pines were growing successfully. The same thing happened with introductions into New Zealand and Africa. Pines require their own mycorrhizae for normal growth.

The resident fungal partners of trees do not all send mushrooms or puffballs above ground seasonally to disperse their spores by air. Some of them never see daylight, living their entire lives beneath the surface. These are the subterranean fungi (hypogeous—fruiting below ground) whose fruiting parts are known as truffles or false truffles.

Fancied by epicureans as a fungal delicacy, truffles form the basis of a commercial industry in Europe and the Mediterranean countries of Africa and Asia. Collectors there train dogs or pigs to locate the fungi by scent.

The truffles in our North Coastal Forests, while as abundant as anyplace on earth, do not excite the human palate, but they play a vital role in the forest food chain. Resembling small potatoes, truffles contain the spores of the next generation within a skinlike case, and as they mature in the soil, give off pungent odors, different with each species—spicy, cheesy, fruity, garlicky, fishy.

The small mammals who sniff them out dig precise little holes just large enough to extract them. As these fungus eaters (mycophagists) garner their underground "potatoes," they become unwitting third partners in a remarkable ecological relationship.

Spruce-hemlock forest plants depend on the subterranean fungi as much as on the aboveground varieties for the mycorrhizal connections that mean prosperity to both. By growing totally in the earth, truffles avoid the quick drying-out of mushrooms and are available much longer in the year.

At the same time, they require animals to disperse their spores, for, unless their spores reach rootlets of conifers to set up a mycorrhizal connection, there will be no future truffles. Subterranean fungi cannot fruit without their coniferous hosts.

The Deer Mice, shrews, and wood rats that consume truffles pass the spores through their digestive tracts intact, blend them in the cecum (blind pouch of the large intestine) with nitrogen-fixing bacteria and yeast, and deposit the whole viable, life-giving mixture in fecal pellets wherever the animals travel. When abundant rains wash the feces into the soil, the spores send out hyphal threads seeking compatible hosts.

Viable spores are commonly spread by earthworms as they bring soil to the surface in castings that are seasonally moved about by rain and wind. Millipedes and ants transport soil, as do swallows, robins, and other birds that use mud as a binder in the nest. Wasps that dig homes in the ground for their eggs often ferry spore-laden earth some distance away.

Douglas Squirrels and Northern Flying Squirrels also act as vectors by digging out truffles and carrying them up trees or stumps to eat, letting some of the spores escape into the air.

The Douglas Squirrels do it by day, but the large-eyed Northern Flying Squirrels feed only at night. For many years wildlife biologists puzzled over why the remains of such dexterous tree gliders as flying squirrels turned up occasionally in the feces of ground-hunting Coyotes and Bobcats. The mystery was solved with the discovery that truffles rate exceptionally high on the squirrels' diet and that after dark they descend regularly to the forest floor to unearth their favorite delicacy. Some, obviously, dally a bit too long.

Most small mammals eat other foods besides fungi, but the California Red-backed Vole (*Clethrionomys californicus*) seems to specialize in them. This chestnut-colored, mouselike rodent lives so exclusively on truffles that it spends nearly all of its life below ground tunneling from one fruiting body to the next.

In the North Coastal Forests where subterranean fungal fruits peak in fall and spring, Red-backs often cache their food for the leaner summer months. They usually bite a hole in the spore case prior to caching, allowing rapid drying of the moist food that might otherwise mold underground. Like flying squirrels, they supplement their menu with lichens during the off-season and, like Deer Mice, they locate their fungi by smell.

Red-backs' dependence on truffles ties them to the coniferous forests where truffles abound. They disappear from clear-cuts and return only when coniferous saplings have reestablished and mycorrhizal fungi have once again become abundant.

The mycorrhizal fungi, which stop fruiting when their coniferous hosts die in clear-cuts or burns, depend heavily on Deer Mice and chipmunks to reintroduce them to open areas. Feeding on fungi in adjacent forests, these small rodents scamper along the edges and cross into the clearings, dropping their fecal pellets loaded with viable mycorrhizal fungus spores, nitrogen-fixing bacteria, and yeast wherever they explore. As conifers reappear, the fungi necessary to form their mycorrhizae are already there.

So the same Deer Mice and chipmunks that sometimes do damage as seed-eaters in reforestation actually help it along in another way, playing vital roles, along with those of all the other wildlife, in the complex dynamics of the North Coastal Forests.

4

Douglas-Fir/
Mixed-Evergreen Forests

Douglas-Fir trees grow widely over western North America. Extremely adaptable, found in more different forests than probably any other western tree species, they follow the Rocky Mountains from British Columbia to central Mexico and the Sierra Nevada to south of Yosemite. Their forests dominate vast areas of western Washington and Oregon and extend southward in California through the Klamath Mountains and Coast Ranges roughly to the Little Sur River in the Santa Lucia Range.

In the California forests, Douglas-Fir grows chiefly allied with mixed evergreen hardwood species of California Bay, Coast Live Oak (*Quercus agrifolia*), Tan Oak, Pacific Madrone, and Canyon Live Oak (*Quercus chrysolepis*), plus deciduous Oregon Oak (*Quercus garryana*) and California Black Oak (*Quercus kelloggii*). These mixed hardwoods form extensive forests on their own in ravines and on moist slopes throughout the southern and central Coast Ranges, southern California mountains, and inland areas. Some of the Coast Live Oaks spread impressively over Coast Range foothills. At Los Osos Oaks State Reserve near San Luis Obispo, their massive limbs writhe and twist outward and upward like intertwined powerful bodies reaching for space.

Inland, on western Sierran slopes, just above the foothills, Interior Live Oaks (*Quercus wizlizenii*) and Canyon Live Oaks often join Pacific Madrone, California Bay and other species in hardwood woodlands. In northwestern California, Douglas-Fir mixes with these hardwood species at the Redwood Forest's eastern edge. To the north, Douglas-Fir/

43

Map 4. Douglas-Fir/Mixed-Evergreen Forest.

Mixed-Evergreen Forests dominate the lower elevations of the Klamath Mountains, with other conifers replacing the hardwoods at higher elevations. To the east, they taper off into drier Oak Woodland and Foothill Woodland as rainfall decreases.

   Much of this Douglas-Fir/Mixed-Evergreen Forest region has gone through a lengthy history of heavy grazing, fire, and logging and presents a complicated mosaic of forests, woodlands, and grasslands. Undisturbed habitats are few, but a fine example of a pristine Douglas-Fir/ Mixed-Evergreen Forest exists in the Northern California Coast Range Preserve near Branscomb, Mendocino County.

   Purchased by the Nature Conservancy in 1959 from Heath and Marjorie Angelo, who had acquired it parcel by parcel over the decades as their wilderness dream, the preserve, originally a little over 3,400 acres, now comprises around 7,500 through a cooperative agreement with the Bureau of Land Management (BLM) to protect the surrounding buffer lands. The BLM addition includes the entire headwaters of Elder

Creek, one of the few remaining undisturbed watersheds in the United States with crystal-clear runoff.

The United States Geological Survey uses Elder Creek as a benchmark for gauging water quality, and the creek is a registered natural history landmark. In 1983 the Coast Range Preserve was designated one of Unesco's Biosphere Reserves of the World, a special recognition of its unique naturalness.

The narrow dirt entrance road to the preserve plunges immediately into forest recesses. Above the sword ferns and huckleberries rise the ruddy shafts of Redwood and the furrowed barks of Douglas-Fir, like giant masts in a forest sea. Smooth-trunked Madrones and lichen-draped Canyon Live Oaks fill the understory. Tan Oaks climb a little higher, halfway up to the canopy. In the wet season every boulder, tree trunk, and limb radiates the green of leafy mosses.

Adjoining the forest, the South Fork of the Eel River, which bisects the preserve, offers many enticing haunts. There are ripples where steelhead break the surface and Common Mergansers (*Mergus merganser*) dive. Bald Eagles (*Haliaeetus leucocephalus*) fly low, scanning for carcasses. River Otters (*Lutra canadensis*) slide down slippery banks. Spotted Sandpipers (*Actitis macularia*) bob on sedge-lined flats. Dippers vanish underwater in their search for aquatic insects. In nearby meadows Black-tailed Deer browse under the round crowns of Oregon Oaks at dawn and dusk.

Wildlife here is often close at hand. The caretaker at the Angelo home awoke one morning to loud growls outside his window and rushed out just in time to see a Mountain Lion bound away with his favorite house cat firmly in its jaws.

A hiker on the Alpine Trail was equally surprised to glance up and meet the eyes of a Mountain Lion resting in a Madrone. With no hesitation, both bolted in opposite directions.

Bobcats, Gray Fox (*Urocyon cinereoargenteus*) and Black Bear roam the trails through the mixed forests. Mink and Raccoons hunt the creeks and rivers. Porcupines come down at night from their daytime retreats in the foliage. The rarely seen Ringtail (*Bassariscus astutus*) and even rarer Fisher (*Martes pennanti*) both leave their tracks. These two fast-moving arboreal predators, along with the Northern Goshawks (*Accipiter gentilis*), keep the Douglas Squirrels eternally on the alert, darting to the under side of a branch at any passing shadow.

The five creeks on the preserve shelter some unique amphibians. The abundant Pacific Giant Salamander lives in all of them. A voracious feeder eating anything from caddisfly larvae to fish, frogs, and other salamanders, it can turn on attackers as large as garter snakes and kill them with its powerful bladelike teeth.

Many of the Giant Salamanders grow to adult size in their aquatic larval form, with external gills and small legs. They may remain larval all their lives. Those that mature hold the record as the largest terrestrial salamanders in the world.

Two small amphibians rate among the rare finds in the preserve. The Olympic Salamander (*Rhyacotriton olympicus*), a 4-inch-long (10 cm) olive and gray recluse with prominent eyes on a small head, lives in a number of the well-shaded creeks, hunting insects, spiders, and crustaceans amid the moss-covered rock rubble. It especially likes places where cold water percolates freely among the rocks and the banks are overgrown with ferns.

Sometimes the smaller and more unusual Tailed Frog shares its habitat. The size of a tree frog with a black line through the eye, a vertical pupil, and a rough olive-brown skin, this tiny amphibian is the only member of its family found outside of New Zealand. Its New Zealand relatives lack the "tail"—not a true tail actually but a pear-shaped copulatory organ which the male uses in internally fertilizing the eggs, a process uncommon in amphibians.

The tadpoles of Tailed Frogs possess large suckerlike mouths with which they cling to rocks in the boulder-strewn creeks of the moist Douglas-Fir forests.

Near the southern limit of their range in the preserve, both of these amphibians associate with old-growth forest, as they do in western Oregon and Washington, needing the stabilizing temperatures, coolness, and shade that such vegetation provides.

The creeks where they live and the adjacent Eel River peak during the winter rains, providing high water for Silver Salmon to swim in from the ocean and spawn in November and December. Steelhead follow in February and March, traditionally jumping Elder Creek's waterfall on Valentine's Day. Dry spells can drop the water so low that the fish stack up in the Eel River at the mouth of Elder Creek in a holding pattern for several weeks until the rains come.

The creeks flow the year around, though summers in the preserve are mostly dry. But I remember one July day when a gentle, steady rain caught two of us hiking in a hardwood grove high on the mountainside. Gray fog closed in quickly, settling around the trees, cutting off visibility across the canyon. Water trickled down the reddish Madrone trunks in long curving streamers, darkening the grooves. Raindrops sent leaves sputtering on the forest floor. Live oaks arching the narrow trail began to swell and drip as the mosses draping their limbs and trunks absorbed moisture. The Tan Oaks, which seemed covered with white blossoms at a blurry distance, turned out to be merely displaying fuzzy new water-flecked leaves. Old leaves farther down the twigs shone like pieces of polished dark green leather.

A hillside of Giant Chain Ferns (*Woodwardia fimbriata*) followed a creek bed into the mist, some of the tall fronds holding individual drops of water that spanned three holes in the leaflets. At several bends of the trail Giant Chinquapin climbed into the overcast, tall and straight with neat, vertically striated bark that blended gray and rose tints to a subtle pink. The only sound—rain falling on leaves—created a world apart.

In another world of its own, a superb old-growth forest of Douglas-Fir thrives on the cooler, moist, north-facing slopes of the preserve. Douglas-Fir trees gain 90 percent of their height in roughly the first 70 years. From then on, through their 600-year life span, they spread mainly in diameter, becoming chunkier in maturity with thick gnarled branches, dense needle spreads, cavities, and broken crowns. All of these niches support a diversity of life that increases with age.

Bruce Bingham and John Sawyer studied seventy Douglas-Fir/ Mixed-Evergreen Forest stands in northwestern California and southwestern Oregon to observe how the distinctive features change in trees from 40 to 400 years of age. They found that by 200 years, these forests have acquired old-growth characteristics: they are strongly two-tiered. Douglas-Fir dominates the upper tier, attaining heights of 250 feet (75 m) or more as widely spaced individuals or in groups. Their branching tops create an open canopy. Hardwoods dominate the lower tier, reaching heights roughly half that of the Douglas-Fir.

Within the forest either tier may become more prevalent from place to place. Forest openings allow regeneration of younger trees so that all sizes and ages eventually share the space. Large snags and downed logs add to the diversity. But "the most diagnostic feature of nonmanipulated old-growth Douglas-Fir/hardwood forests is the presence of large old trees in densities adequate to form an upper tier that dominates over a hardwood layer" (Bingham et al. 1991:376).

These treetop canopies contain their own plant and animal life. Botanists using mountain climbing gear on a 450-year-old Douglas-Fir in western Oregon discovered that the higher they went the more kinds of organisms they found. The treetop plant communities were chiefly lichens, symbiotic combinations of algae, which have chlorophyll and make food, and fungi, which absorb water and give support. They were prolific—over 120 species identified in the old trees studied.

They ranged from lichens that pioneered an existence on young twigs to complex communities on more mature branches. Some species favored the more rained-on side of trunks, others the drier. Flat, leaflike lichens occupied the upper side of canopy branches, liverworts the lower side.

Some lichens, such as *Lobaria oregona,* are well known for taking nitrogen from the air and releasing it to the forest floor when they fall, adding an element needed by all life. Many of them provide food for other creatures of the high trees—Red Tree Voles, flying squirrels, insects, nematodes, mites. As many as 1,500 species of invertebrates often live in a single stand, including many kinds of predaceous spiders, leafhoppers, butterflies, moths, flies, and beetles.

Among the mammals, none has a closer association with the old-growth forests than the Red Tree Vole. This small brick-red mouse, with well-concealed ears and a long black furry tail, spends almost its entire life high in old Douglas-Fir, rarely descending to the ground. Its bulky nest, a foot or more wide, built out on a branch or next to the trunk, consists of twigs, branches, and shredded bark from its host. After the vole dines on the fleshy part of the Douglas-Fir needles, it uses the leftover resin ducts of the midribs to line the nest. Seldom seen in its arboreal home, it turns up frequently in the pellets of its chief predator, the Northern Spotted Owl (*Strix occidentalis caurina*).

Northern Spotted Owls live secretively deep in the old-growth Douglas-Fir forests. Sleepy by day, they come alive at dusk and tune their acute nocturnal vision and phenomenal hearing to the most minute sight or sound. They dive silently from elevated perches onto their prey—Red Tree Voles, Dusky-footed Wood Rats, Saw-whet Owls (*Aegolius acadicus*), and others. With facial feathers that form the equivalent of a satellite dish funneling sound to the ears, and a right ear larger than the left to help triangulate the sound, they can catch their victims through hearing alone.

The canopy of the old-growth forests protects the Spotted Owls somewhat from their enemies, Northern Goshawks and Great Horned Owls (*Bubo virginianus*). The snags and dead standing trees with holes for flying squirrels, woodpeckers, the tiny Northern Pygmy Owl (*Glaucidium gnoma*), and Western Screech Owls (*Otus kennicottii*) provide a steady food supply.

Northern Spotted Owls do not build nests of their own. They utilize suitable broken tops of tall Douglas-Fir trees, old Goshawk nests, or tree cavities for raising young. And although pairs do not nest every year, they stay together for many years, perhaps for life, reforming the pair bond annually. Gentle and trusting in the presence of owl watchers, they will allow an approach of a few feet and often doze off while observed.

Cameron Barrows studied Spotted Owls, both at the Northern California Coast Range Preserve and elsewhere in California. He observed their explicit reactions to heat, to which they are extremely vulnerable. They show heat stress signs at 85° F (29° C), 15° F lower than Great Horned Owls. Consequently, in summer the Spotted Owls roost in heavy shade on north-facing slopes. They choose the coolest place around, often near a creek. As temperatures rise on hot days, they move lower into Tan Oaks and dogwoods: midday temperatures at the forest floor range from 5 to 9° F (3 to 5° C) cooler than the canopy.

If warmth gets to them, they raise their toes to radiate heat through blood vessels on the bottom of the feet. They begin a gular flutter, passing air rapidly over the moist membranes of the throat. They spread drooping wings out from the body and elevate dorsal contour feathers. And if air temperatures approach 90° F (32° C), they dunk themselves in the creek, drenching face and breast thoroughly.

Spotted Owls occur in scattered areas from southern British Columbia to the Sierra Madre of Mexico, and from western Washington south through forests of California. Where the forests do not have old-growth characteristics, they occasionally contain some darker, cooler microclimates, often with northern exposures and adjacent water which the owls can use.

The Northern Spotted Owl's home range requirements of 2,400 to 7,800 acres of suitable habitat per pair are met especially favorably within the ancient Douglas-Fir Forests from western Washington south through the Northern California Coast Range Preserve. The fate of the northern subspecies of Spotted Owl in these areas revolves inextricably around the survival of the equally threatened, old-growth forests.

Bark Beetles

The logs and snags that house much of the wildlife of ancient Douglas-Fir Forests represent, in good measure, the work of boring insects. California harbors over 170 species of bark beetles alone, ranked among the most destructive insects in the West. Small and seldom seen, they kill more trees than forest fires. Nearly all conifers are subject to them, especially in times of drought, and Douglas-Fir is no exception. Pitch tubes on the bark, piles of sawdust at the base of trunks, and browning needles commonly indicate the beetles' presence.

Douglas-Fir Beetles (*Dendroctonus pseudotsugae*) play a major role in the development of both snag and log habitats in their forests. Each spring thousands of female Douglas-Fir Beetles leave their winter homes and take wing in search of new hosts. Finding a Douglas-Fir weakened by drought, wind damage, fire, or disease, they bore into its bark. Though only 1/4 inch long (6 mm), these dark brown beetles with reddish wing covers overwhelm a sick tree's defenses by their sheer numbers.

As each female reaches the inner bark (phloem), she starts an egg gallery. On her body she carries the Blue-stain Fungus. The fungus quickly spreads into adjacent water-conducting channels (tracheids) and blocks them. These channels normally provide the pressure required to pump resin, the chief chemical defense of Douglas-Fir. When resin is available, it can flood the beetles' galleries and drown them.

With the tracheid pressure cut off by the fungus, resin cannot pour into the galleries. The beetles live on, laying eggs that turn into larvae, pupae, and adults. Eventually the tree dies from water blockage by the fungus and phloem loss to the beetles.

With thousands of beetles boring in the same bark layer of a weakening Douglas-Fir at the same time, how do they keep from invading each other's galleries? Julius Rudinsky and colleagues at Oregon State University determined to find out. The beetles, they discovered, employ a very neat acoustic system.

The females make a single or double clicking sound as they tunnel, clicking intermittently. But if a neighbor comes within 2 inches (5 cm) they click continuously, signaling clearly "this territory is occupied." The clicks also tell other females exploring the outside bark for entrance sites that this area is already taken, but they welcome male beetles, both with the sounds and with chemical signals called pheromones.

The whole life cycle of the Douglas-Fir Beetle is regulated by the combined effects of acoustic signals—female clicks and male chirps—along with the pheromones given off by both.

If two or more male beetles arrive at a female's entry hole at the same time, drawn by her clicks and pheromones inside the tunnel, they crash into each other like "two tanks," fighting violently, emitting loud "rivalry chirps." As Lee Ryker, one of the research team, described it, these are very different from the "attractant chirps" males use on other occasions (Ryker 1984:119). The first arrival usually eventually gains entrance.

The male beetle's courtship is anything but friendly. He bites and jostles the female, who defends herself by blocking the tunnel with her heavily armored posterior. Gradually the male softens his attacks, modulates his chirps to a courtship tone, and begins to stroke her wing covers with his forelegs. She responds speedily to this milder courting, and the pair usually mate within a few minutes.

The vigorous courtship probably serves a useful adaptive purpose. To survive it, both must be robust and in top shape to pass on their traits to a new generation of beetles.

Such intimate knowledge of the reproductive behavior of Douglas-Fir Beetles came only with decades of patient and innovative research. Using a square of fresh Douglas-Fir bark clamped between two sheets of plexiglass, with beetles inside, the scientists attached a microphone to hear and record the sounds. An oscilloscope analyzed the sounds. Gas chromatography and mass spectrometry analyzed the pheromones.

Fallen Trees

Fallen trees, in various stages of decomposition, often cover up to 20 percent of the forest floor in ancient Douglas-Fir Forests. Whether they were felled by beetles, killed by fire, blown down by the wind, or toppled from old age and decay, all take on a whole new valuable role in the forest ecosystem.

Chris Maser and James Trappe collaborated in ferreting out the succession of life in a fallen, ancient Douglas-Fir.

How and where each tree falls determines much about its fate. If one happens to land along the contour of a slope, it immediately becomes a natural erosion barrier for everything tumbling down from above. New cover for small tunneling animals quickly piles up along it. The side contacting the ground absorbs and retains the most moisture, keeping that area wet even through the summer drought. The downslope side provides shelter for larger animals.

As the fallen tree decays over the decades, it offers continuously changing external and internal habitats to an array of plants and animals. The top, with the smallest diameter, decomposes fastest.

The trunk, made up of a number of different tissues, decays at each

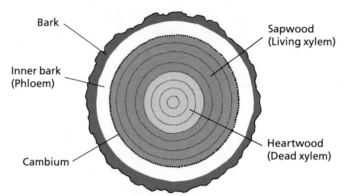

Figure 16. Cross-section of tree trunk. Vascular cambium is the
cell-producing layer of the tree. The inner bark
transports food throughout the plant. Bark insulates
and protects from injury. Sapwood transports and
stores water and dissolved minerals. Heartwood
makes up the dead, structurally supportive center.

tissue's own rate. The bark hangs on for years as a wall that invaders
must penetrate to get inside. Just under the bark lie the most physio-
logically active layers of the tree—the inner bark (phloem) that transports food in life; next to it, the growing layer of the tree (cambium);
and inward from the cambium, the sapwood. Sapwood (outer xylem)
is the mostly living portion of the pipeline that transports and stores
water and dissolved mineral salts in life. The entire center of the trunk
is filled with heartwood (dead inner xylem).

In a fallen tree, the inner bark and cambium, full of the most nutritious, high quality food, are promptly invaded by bark beetles. The first
of a procession of selectively timed arrivals, the beetles must be speedy
if they are to complete their entire life cycles in these perishable layers
before the tissues dry out.

The female Douglas-Fir Beetle chews through the outer bark in early
spring into the inner bark and cambium, where she tunnels a 2-foot-
long gallery (60 cm). Here she lays her eggs. The larvae that emerge
chew their own feeding galleries through the inner bark, packing them
with their refuse (frass) as they go. They overwinter as mature larvae
and emerge as adults the next spring.

As in the standing trees, with these first penetrations of the log's
interior, the beetles transport into the galleries the fungal spores that
commence their own breakdown of the wood. Other wood-boring
beetles join the feast. Brilliantly colored Golden Buprestids (*Buprestis
aurulenta*), metallic green with coppery wing margins, deposit eggs in

log crevices. Their larvae bore tunnels up to 15 feet (5 m), packing them with their frass. In good quality wood, the larvae turn into pupae and emerge as needle-feeding adults in 2 to 4 years. In poor habitat, it may take 10 years or longer.

Ambrosia bark beetles do not eat the wood they invade but leave their castings on the bark as light-colored powder. Their lives depend on the Ambrosia fungi they bring into their galleries. As the fungi grow, the beetles use them as food. Very selective about the flavor and texture of the fungi, each species of Ambrosia beetle gathers its own particular kinds. If moisture in the galleries drops too low, the entire fungal crop may fail and the beetles starve; if the galleries stay too wet, the fungi grow so expansively that they smother the beetles in their own dens.

Once fungi enter the tree through beetle tunnels, they remain with it in one form or another all the way to the finish, their lives intertwined with the insects and invertebrates that feed on them and disperse their spores.

Carpenter ants (*Camponotus* spp.) often become the first insects in a down trunk to use it as a nest and to commute in and out regularly. Anyone who has ever sat on a log in the woods for 10 minutes knows these large 3/4-inch (2 cm) brownish black ants.

The young male and female carpenter ants leave the nest in early spring on mating flights, often filling the air as they emerge simultaneously from scattered nests. The males die after mating, but the young mated queens either enter established colonies to replace "burned-out" queens or search for small cavities in fallen trees where they can start new colonies.

Enclosing herself in a brood cell, each queen breaks off her wings and lays a few eggs. When the eggs hatch in about 10 days, she feeds the larvae secretions from her salivary glands. In a month, when the larvae mature as workers, they take over the construction of the nest. They cut parallel galleries through the soft, decaying wood, and drop the sawdust outside.

Carpenter ants do not eat wood. They forage around the log for caterpillars, dead insects, and, primarily, the sweet honeydew secreted by aphids. Sometimes they bring aphids into their nest in winter and carry them back out onto plants in the spring, tending them carefully. The workers take care of the rapidly increasing eggs and larvae as the nest grows, feeding the queen and doing all the necessary maintenance chores.

Carpenter ants require a moist environment. So do termites and

Figure 17. Old-growth Douglas-Fir Forest. The Red Tree Vole (above top) lives its entire life high in Douglas-Fir branches. The Red Crossbill (above bottom) uses its crossed beak effectively in removing seeds from Douglas-Fir cones. The Spotted Owl (opposite top) preys on the smaller Pygmy Owl and Northern Flying Squirrel (opposite). Chain Ferns thrive on the forest floor beneath Poison Oak vines that climb the furrowed Douglas-Fir trunks. *Large inset*: Douglas-Fir Beetles bore into inner bark. *Small inset*: beetles start egg galleries.

most of the log's growing number of inhabitants. Fallen trees have no trouble supplying this. The bark keeps moisture in during the early years. The tunneling of the wood-boring insects spreads microbes which break down cell walls with their enzymes. The breakdown process produces carbon dioxide and water, further moistening the interior and encouraging decay.

When the wood is just right for Pacific Dampwood Termites (*Zootermopsis angusticollis*), the odors of certain acids and aldehydes produced by particular fungi attract them to the tree. Winged termites, like ants, appear out of logs in vast numbers at the mating season, flying short distances before dropping to earth and shedding their wings. Courting pairs walk in tandem to a fallen tree, the female leading. They excavate a small chamber, mate, and start a new colony.

The queen turns into an egg-laying machine, with all the early termites hatching into nymphs and sterile soldiers. Some of the nymphs become immature reproductives that take on the work of keeping up the nest. The soldiers, with enlarged jaws, provide defense.

Meanwhile the termites eat themselves through galleries in the interior of the tree. They can digest the wood and get energy from it only because of a unique three-way process involving themselves and certain one-celled protists and nitrogen-fixing bacteria that live in their guts. The termites eat the rotting wood; the protists digest the cellulose in the wood and convert it into a usable termite food. The termites, in turn, provide wood-particle foods for the protists; the nitrogen-fixing bacteria produce the nitrogen required whenever the termite's nitrogen intake is deficient.

Nitrogen figures prominently all the way through a fallen tree's cycle. Originally absorbed by the roots of the tree and incorporated into the wood, nitrogen in the down log gradually moves from one life form to another as decomposition frees it. Fungi and bacteria add it to their own structure when dissolving tissues. They, in turn, pass it on when eaten by mites, springtails, and beetles.

These fall victim to the pseudoscorpions that Maser and Trappe describe walking "upside down on the ceilings of small caverns within rotten wood" or sitting in their silken nests, pinchers ready to grab unsuspecting passersby (1984:33).

The moist inner trunk channels grow ever more perilous for insect tenants as the tree becomes increasingly riddled. Spiders lie in wait. The Pacific Folding-door Spider (*Antrodiaetus pacificus*), largest of its kind living in fallen trees, builds a silk-lined tube from which to seize prey. It can close the tube by pulling the rim on either side toward the middle, like a folding door.

Spiders, too, have enemies. Centipedes paralyze them with a bite from poisonous jaws. From the first invasion of a fallen Douglas-Fir on through its continuously growing chain of life, each insect and invertebrate within the log has its predators and parasites.

Many of them are eventually eaten by the California Slender Salamander (*Batrachoseps attenuatus*) and the Clouded Salamander (*Aneides ferreus*). The Slender Salamander preys especially on springtails and mites, its pencil-slim body the perfect size to slither through the narrow confines of termite and beetle galleries.

The Clouded Salamander particularly favors down trees in openings in the forest. The female lays eggs in late spring or early summer in cavities in the rotten wood and stands guard over them as they hang on stalks from the ceiling or wall of her nest chamber. Her diet, as well as that of her young, leans heavily to ants in the summer and a mixture of isopods, beetles, earwigs, mites, springtails, flies, centipedes, and termites the rest of the year.

Competing for much of the same food are the shrews and Shrew-moles. The Trowbridge Shrew (*Sorex trowbridgei*), with minute, poor eyes but acute senses of smell, hearing, and touch, especially relishes down trees and is known to eat at least forty-seven kinds of prey that live there, among them centipedes, spiders, slugs, snails, and beetles.

The Shrew-mole, beautifully adapted for digging under and alongside fallen trees with its stout front claws, also uses its hidden ears and tiny eyes effectively in hunting the tunnels.

In old-growth forests all the essential minerals needed to maintain a healthy flora and fauna recycle naturally over the centuries, linking life and death interdependently. A living tree draws what elements it needs from the soil and air and incorporates them into its woody cells, giving off oxygen to its surroundings. The fallen tree gradually returns these elements to the earth for use by succeeding generations and, in the process, feeds and shelters its current residents.

As research by Jerry Franklin and colleagues has attested, a single old-growth Douglas-Fir can visibly affect its ancient forest site and the flora and fauna around it for nearly 1,000 years. Living, it influences its surroundings for 6 centuries. Fallen intact, it disintegrates gradually over 300 or more years, a prime home and food source for a stream of inhabitants. And it remains a useful component of the soil long after it shrinks into a mound, a fertile nursery for new young plants.

5

Closed-Cone Pines
and Cypresses

Highway 1 curves down the northern and central California coast through scenic, intermittent forests of one of the state's most variable conifers, Bishop Pine (*Pinus muricata*). Since its discovery in 1830 by Irish botanist Thomas Coulter in the vicinity of San Luis Obispo, Bishop Pine has become known, along with Monterey Pine (*Pinus radiata*) and Monterey Cypress (*Cupressus macrocarpa*), as one of the more picturesque members of the state's Closed-cone Pine and Cypress communities.

Four pines and ten cypresses make up this ecological category, Once more widespread in ancient forests, they were reduced to small stands as the climate dried up over the past 10,000 years. Today they form unique, often isolated arboreal "islands" sprinkled over the length of California's coast and inland mountains. Most eke out their existence on generally infertile soils, some restricted to serpentine. All have cones that require fire or intense sun to open. Some delay their opening for years (serotiny).

Bishop Pines spread over one of the longer territories in the group, ranging discontinuously from Trinidad Head in Humboldt County to Lower California and to Santa Cruz, Santa Rosa, and Cedros Islands. Intensely individualistic, Bishop Pines adapt to their thin coastal belt in whatever ways it takes to survive. On the exposed northern headlands and bluffs above Pacific Ocean surf, their multiple trunks twist and contort to stay low and alive in the battering winds and to keep their two-needled green tufts intact on the leeward side.

Map 5. Closed-cone Pines and Cypresses.

With favorable soil moisture in forest settings, these pines grow tall and straight, their rounded tops interlocking. Along forest edges where there is open space, the trees branch freely, bending at right angles into all sorts of striking formations. In northern coastal state parks such as Van Damme and Salt Point, Bishop Pines share tall, shady forests with Douglas-Fir, Grand Fir, California Bay, Tan Oak, and Pacific Madrone.

Along the northern and central California coast, Monterey Pines have been planted in many places where they and their natural seedlings now intermingle with the Bishop Pines. Both grow side by side in Jughandle State Reserve near Fort Bragg and at Gualala County Park.

Although at first glance the two species show needles about the same length, 4 to 6 inches (10 to 15 cm), a needle count will reveal two per bundle for Bishop Pine and three for Monterey. The Monterey's foliage has a distinctly bright yellow-green color when compared to the Bishop's blue-green. Both display closed, oblique cones that clasp and often encircle the branches. However, Bishop Pine's are smaller with stout spurs on the cone scales; Monterey Pine's are larger and rounded, with tiny cone scale prickles that eventually weather off.

Although Monterey Pine's native range today is restricted to three mainland areas from Año Nuevo south to Cambria, fossil records indicate that the species once occurred as far north as Point Reyes and shared a common ancestor with Bishop Pine during the Pleistocene epoch about one million years ago.

Point Reyes National Seashore furnishes the current setting for some of the most extensive and beautiful groves of Bishop Pines along the coast. On the granite soils of the north end of Inverness Ridge, they climb the leeward side from Tomales Bay in rich, luxuriant forests. Intermingled with the tall arches of bay trees, occasional red-barked Madrones and silver-trunked Coast Live Oaks, the Bishop Pines' dome-shaped crowns disappear into the summer fogs like scenes from Oriental landscapes. Lace Lichens (*Ramalina menziesii*) drape the gray vertically furrowed trunks and branches with soft ghostly streamers.

The understory carries many of the berry-laden shrubs of the North Coastal Forests—Salal, huckleberries, Poison Oak, Wax Myrtle. Their fruits attract the birds of the Bishop Pine Forests.

In summer the upscale fluting song of Swainson's Thrush sounds everywhere, though the slim brownish thrush itself may be elusive. Pygmy Nuthatches (*Sitta pygmaea*) chip out their own nests in snags and pipe a Morse-code-like "dit dit dit" as they poke gray brown-capped heads into the spaces around Bishop Pine cone whorls in search of food. The high lisp of a Brown Creeper (*Certhia americana*) accompanies its climb up a Bishop Pine trunk, its slender decurved bill probing the deep furrows for insects and spiders.

Some of those spider webs end up in the nests of Bushtits (*Psaltriparus minimus*). Living in loose flocks during the nonbreeding season, Bushtits forage through the forests, keeping in touch with gentle high "tsee-tsee-tsee" notes, as they glean small insects from the underside of leaves and twigs.

Only slightly larger than hummingbirds, these tiny gray birds build distinctive gourd-shaped nests, seemingly disproportionately spacious for their seven or eight dainty white eggs. They suspend these long soft pouches from twigs 15 feet or more (4.5 m) above the ground and decorate the intricate collection of woven mosses, lichens, willow down and webs with moth wings or flower blossoms, often spending 6 weeks in construction.

Two females and one male, or two males and one female, commonly share duties at the nest, flying secretively in and out of the opening high on one side. When the young leave, they vacate all at once, if the ever watchful Steller's Jays haven't discovered the nest, torn it apart and eaten the fledglings.

The trill of Dark-eyed Juncos (*Junco hyemalis*), the ring of Song Sparrows, and the dry chatter of the black-topped Wilson's Warbler (*Wilsonia pusilla*) are common sounds in the Point Reyes Bishop Pine forests. Purple Finches (*Carpodacus purpureus*) fill the air with rapid warbles; Olive-sided Flycatchers (*Contopus borealis*) call loud "Hick, three beers" from treetops. Overhead, at any time, may sound the clear, liquid whistles of Ospreys (*Pandion haliaetus*). Plunging feet first into Tomales Bay to snatch unwary fish, they bring the catch home to hungry young in crowns of nearby Bishop Pines.

The steep climb from this rich forest setting to the barren granitic hogbacks high on Inverness Ridge traverses an almost solid California Bay woodland, strong with the scent of crushed bay leaves. As Bishop Pines mount the slope, shrinking increasingly in size, they pass through a Chaparral belt where the shrubby form of the Giant Chinquapin tree predominates. The Golden Chinquapin bush (*Chrysolepis chrysophylla* var. *minor*) fits its name. Its new leaves emerge a bright golden; the older pointed leaves, folding up at the midrib, show green upper sides dotted with hundreds of tiny gold scales and lower sides solid golden.

On the thin deficient soil of the ridgetop hogbacks, Bishop Pines form a dwarf forest of stunted trees growing only a few feet high but producing mature cones. A rare variety of Huckleberry Manzanita (*Arctostaphylos glandulosa*), with fine, densely hairy twigs, spreads mat-like between the trees.

Up and down the coast, Bishop Pine's adaptability shows itself in other natural "bonsai" forests, including the best known Pygmy Forest of Van Damme State Park near Little River. Here a one-third of a mile (.5 km) boardwalk trail winds through one of the most unusual forests in the temperate zone—an elfin forest of gray, twisted trees, growing on the leached, acid soil of the Mendocino plains.

Mendocino or Pygmy Cypresses (*Cupressus goveniana* ssp. *pigmaea*), standing 1 to 5 feet high (.5 m), fill dense canelike thickets, their slender, silvery stems encrusted with lichens. Many of the trees are 40 years old with trunks one inch wide (2.5 cm). Some cypresses only a foot high show sixty or more annual rings. Yet these pygmies are biologically mature, bearing viable seeds inside their round, closed cones.

Figure 18. The Pygmy Forest of the Mendocino Coast hosts a
strange world of elfin trees, Mendocino Cypress
(upper left), Bolander's Pine (center), and Bishop Pine
(right). Each may take half a century to attain dwarf
stature in the deficient podzol soil. California Rose-
Bay (lower right) and Evergreen Huckleberry (lower
left) provide spring color.

Equally stunted Bolander and Bishop Pines mingle with them. Bolander Pine (*Pinus contorta* ssp. *bolanderi*), a highly localized closed-
cone form of the more widespread Lodgepole Pine–Shore Pine complex, possesses the usual two short needles and small cones of the group.
Its serotinous cones remain closed on the tree's crowded, spindly
canes for many years. Shore Pines (*Pinus contorta* ssp. *contorta*), small
trees with dense rounded crowns, grow scattered along sand dunes and
coastal bluffs from Mendocino County northward to Alaska and are
considered closed-cone species by some botanists.

Bolander Pines fight for survival in the same slow-growth pattern as Mendocino Cypress, pushing up 1 to 2 feet over 40 to 50 years. Bishop Pines, which reach 80-foot (24 m) heights elsewhere, here, over decades, are lucky to make 8 feet (2.4 m).

The explanation for all this pygmy growth lies in a remarkable interaction of coastal geology with heavy regional rainfall, special plants and unusual soil. Over the past half million years the Pacific Ocean carved a series of marine terraces into the Mendocino coast, each terrace about 100,000 years older than the one beneath it and 100 feet (30 m) higher. Along most coasts such terraces erode back into nonexistence, but at Jughandle State Reserve, a few miles north of Van Damme's Pygmy Forest, they have remained largely intact. They tell a unique story.

The first terrace, the most recent one, only 100,000 years old, is a Coastal Prairie today, colorful with grasses and wildflowers—rosy sidalceas, lupines, yellow composites—all tolerant of the salty ocean spray that blows continually onshore.

A trail climbs the "giant staircase" to the second terrace, farther inland and another 100,000 years older. Cleansed of salt over time by the heavy regional rains, the soil on this terrace supports a mixed, mostly coniferous forest of Sitka Spruce, Grand Fir, Douglas-Fir, Western Hemlock, Redwood, Bishop Pine, and California Bay. Its shaded forest floor carries plants typical of the Redwood and North Coastal Forests—sword ferns, trilliums, and Redwood Sorrel.

But even in the richness of this forest there are signs of change. The heavy winter rains drain poorly from the flat terrace, so the trees must tolerate standing water at times. Here and there patches of the grayish soil known as podzol replace the more fertile darker kind, indicating where vital nutrients have been consumed by conifers and leached out by rainwater.

The soil has become so impoverished by the slope to the third terrace that the mixed conifers can no longer live in it; they are succeeded by Bishop Pines which can survive on a more spartan supply of nutrients.

Even the Bishops become stunted on the upper oldest terraces. These leached soils of 300,000 to 500,000 years, severely deficient in calcium, potassium, magnesium, and phosphorus, can support only minimal dwarfed plant life. This is pygmy forest habitat. The plants that live here all tolerate highly acid soil which their acid-rich needles and leaves continually augment. Theirs is probably the most acid soil in the world, with a pH of 2.8 to 3.9 on a scale where normal soil ranges from 6 to 7.5.

A cross-section of typical pygmy forest soil would show, first, a very thin layer of humus, then a foot (.3 m) or so of leached gray podzol, and finally a layer of reddish brown hardpan. The acid-tolerant plants themselves help to build the hardpan by producing chemicals that leach iron from the surface and carry it below where it mixes with quartz grains to form an impenetrable "cement."

Thus the plant roots lie imprisoned in 12 inches (30 cm) of impoverished soil above the hardpan which they cannot penetrate. And during the heavy winter rains, coffee-brown water pools soak the roots in an acidic bath.

Despite these stringent conditions, the diminutive forests sport some colorful acid-loving shrubs among the pines and cypress. California Rose-bay unfolds striking pink and white flower clusters in May. Western Labrador-tea's white blossoms (*Ledum glandulosum* ssp. *columbianum*) and Salal's urn-shaped pinkish white bells brighten the gray surroundings. California Huckleberry turns crimson with new leaves in May and June, and the low-growing Fort Bragg Manzanita (*Arctostaphylos nummularia*) and taller Hairy Manzanita (*Arctostaphylos columbiana*) add rich reddish stems year round.

Here and there a lone Bishop Pine towers above the Lilliputian forest. Hans Jenny, who, along with his wife Jean, is largely responsible for the recognition and preservation of pygmy forests in northern California, excavated one taller Bishop Pine and discovered the reason for its increased growth. Its enormous taproot had somehow penetrated the hardpan and extended into the more fertile strata below.

Mendocino Cypress grows to 100-foot heights (30 m) and 3-foot diameters (.9 m) on better drained soils. One of the taller California cypresses, it sometimes invades the margins of adjacent Redwood Forests following fire or logging. But Bolander Pine, which is restricted to the Mendocino pygmy forests, rarely does either.

Wildlife finds slim pickings among these miniplants. Chipmunks, Steller's Jays, and other birds discover the berries. Douglas Squirrels and Brush Rabbits occasionally explore for seeds. A number of the more than 200 species of buprestid beetles in California find homes here. The adult beetles generally feature bright metallic colors and streamlined bodies, and commonly feed on pollen. Both the Golden Buprestid, with copper-edged green wings, and the Western Cedar Borer (*Trachykele blondeli*), with golden reflections on shiny green wings, produce larvae that mine the pines and cypress.

But this space-rich forest floor is basically species-poor. Invading seeds meet a hostile, deadly reception. Old plants have to die to release the organic matter and nutrients which new ones need to get started. There is no surplus, and the successors must possess the same acid-loving genes as their progenitors.

So the pygmy forest ecosystem is self-sustaining—a strange but true climax forest. Hans Jenny has written that "it comes as close to a terminal steady-state system with balanced inputs and outputs as can be expected to be found in nature" (Jenny 1969:73).

The Mendocino pygmy forests are restricted to a roughly 18-mile-long (30 km), narrow coastal strip between Fort Bragg and the Navarro River. Further south, on a more obscure marine terrace in Monterey County, old beach deposits host a different pygmy forest. Another scarce survivor from the more widespread closed-cone forests of the past, this remnant group of Bishop Pines and rare Gowen Cypress (*Cupressus goveniana* ssp. *goveniana*) exists in the sterile, acidic, podsolized soil of Huckleberry Hill in the Del Monte Forest.

Gowen Cypress is found only here and in the Gibson Creek grove of nearby Point Lobos State Reserve. Gowen, along with stunted Bishop Pines, grows on the very worst soil. On adjoining better soil, the short, broad cypresses drop out and Bishop Pines grow taller. Monterey Pines take over on surrounding normal soils.

Huckleberry Hill seems to be the only place where Monterey and Bishop Pines survived the climatic changes of the millenia together. Though closely related, they remain genetically distinct. Their pattern of shedding pollen months apart undoubtedly helps prevent most hybridization.

The Pebble Beach fire of 1987 wiped out a section of the Huckleberry Hill Bishop Pine–Gowen Cypress forest. True to form, a year and a half later, both species were revegetating the spaces between the blackened trees with seedlings a foot or more (30 cm) high. Bear Grass (*Xerophyllum tenax*), near its southern limits here, and California Huckleberry were also vigorously resprouting.

Fire melts the resin seals on Bishop Pine's tightly closed cones, releasing the seeds and, at the same time, preparing a mineral-rich bed of ash. Fire dries out the cypress cones as it dissolves their resin so that they open and drop their seeds.

Bishop Pine cones will sometimes open audibly on hot September days, making a crackling sound easily heard by anyone in the vicinity. But the even-aged stands characteristic of most of their forests unquestionably owe their existence to fire. Unless rejuvenated by flames,

Bishop Pines succumb in less than a century to fungal, squirrel, and beetle damage and the debilitations of old age.

Monterey Pines also bear cones that will open after two or three days of intense autumn heat and drop off within a few years, but they, too, depend on fire for long-term rehabilitation of their stands. They occur today natively in only three places on the mainland: at Año Nuevo 30 miles (48 km) north of Monterey; at Cambria 65 miles (104 km) south; and, most extensively, on the Monterey Peninsula.

Although not plentiful in California, they rank as one of the most widely planted lumber trees in the world. They fill plantations in Chile, Australia, and Spain, cover hillsides in New Zealand, and grow abundantly in South Africa, Kenya, Argentina, and Uruguay.

In California's Point Lobos State Reserve, Monterey Pine forests vary from open-spaced woodlands where overhead light supports low foliage, grasses, and wildflowers to continuous, tall shafts of trees topped by crowns of rich green needles. At meadow and Chaparral margins, they furnish scenic backdrops and valuable forest-edge habitat for wildlife.

In the more mature forests, Chestnut-backed Chickadees and Pygmy Nuthatches make good use of the rotted pine stumps. The Chickadees nest in cavities 4 to 10 feet (1 to 3 m) above the ground, flying into well-screened nests quickly and silently. Nesting at this low level usually frees them from the noisy, fiercely competitive nuthatches who prefer holes 30 to 60 feet up (9 to 18 m) and dig their own.

California Quail (*Callipepla californica*), Steller's Jays, and Anna's Hummingbirds (*Calypte anna*) are other full-time residents. The mild winter climate brings an irregular influx of Evening Grosbeaks, Red-breasted Nuthatches (*Sitta canadensis*), and Yellow-rumped Warblers. Summer draws Allen's (*Selasphorus sasin*) and Rufous Hummingbirds (*Selasphorus rufus*) and all the species of adjacent oak woodlands and streamside thickets.

It is also the season when the Silver-spotted Tiger Moth (*Halisidota argentata*) sails among the Monterey Pine branches at twilight. With silver spots prominent on its reddish brown forewings, and a 2-inch wingspread (50 mm), the moth deposits clusters of green eggs on needles and twigs.

The hairy brown caterpillars that emerge feed on pine foliage, overwinter on the tree in loose webs, and recommense eating needles in the spring. When two-thirds grown, they move out on their own and mature into reddish brown caterpillars with tufts of yellow, brown, and black hairs that can cause a rash on sensitive human skin. The silky brown cocoons that the caterpillars spin end up attached to forest floor wood or to the stripped pines.

Some cocoons are robbed of their juicy contents by Western Fence Lizards (*Sceloporus occidentalis*), Slender Salamanders, or the Deer Mice and Meadow Voles (*Microtus* spp.) that scurry about the brushy woods. Western Gray Squirrels (*Sciurus griseus*) depend heavily upon the Monterey Pines for food, carrying off freshly cut cones, along with acorns from the Coast Live Oaks. Black-tailed Deer and Raccoons frequent the forests.

All of this woodland wildlife can be found in other California habitats as well, but seldom in so picturesque a setting dominated by one of the rare pines of the world.

An even more dramatic forest exists close by. Where the outermost promontory of Point Lobos juts headlong into the sea, it contains one of the two remaining stands on the planet of Monterey Cypress in its native habitat. The other stand lies a few miles north on Seventeen Mile Drive.

One of the rarest conifers in North America, Monterey Cypress is another relic from the past. During Pleistocene times the tree grew along the Pacific Coast south to Mexico. As the climate changed, it gradually disappeared from all except these two groves on the north and south side of Carmel Bay.

Embracing fogs and ocean spray, the cypresses are seldom found more than a few hundred feet from the pounding surf. Buffeted by the winds, their silvery trunks cling to sheer cliffs directly above the waves. Some crawl along the rock face, blending with the gray boulders. Wind and sea have blasted many into splintered remnants.

A little farther back from shore, the cypresses form forests of more heavily foliaged trees, merging with the battered ones of the exposed bluffs. These are eerie woods when the cold, wet fog rolls in, swallowing everything except the dark trunks and the long gray beards of Lace Lichen dangling from black branches. The only bits of color breaking the somber mood are the velvety films of red-pigmented green algae (*Trentepohlia aurea*) blanketing many of the gray barks.

But when the sun shines through the silvery trunks and green foliage against a background of dark blue water, billowing breakers and bright blue sky, this Monterey Cypress forest radiates a color harmony rare in the natural world.

Few birds or mammals find the tangle of branches and dense foliage of the cypress favorable living quarters. Wood rats often build bulky nests among fallen debris. Juncos forage on the ground; Winter Wrens search among the thicker branches; Townsend's Warblers (*Dendroica townsendi*) prefer the cypress over the nearby pines; and House Finches (*Carpodacus mexicanus*) build nests of lichen that are almost indistinguishable from lichen-covered limbs.

Monterey Cypress, so rare in the wild, has like Monterey Pine, been planted successfully in many parts of the world, including New Zealand, South America, and the California coast. Considered one of the hardier trees, valued for its beauty as timber and as a windbreak, ironically, this cypress, when planted in California away from the coast, usually succumbs to Cypress Canker (*Coryneum cardinale*). Salt spray apparently keeps the canker's fungal spores under control.

Seven other species of cypress native to California live mostly inland on islands of poor soil. Their scattered woodlands vary in size from a few trees to a few thousand, often in groves isolated from others of their own kind.

McNab Cypress (*Cupressus macnabiana*) and Sargent Cypress (*Cupressus sargentii*) range over the larger areas of the group, sometimes occurring together in the northern Coast Ranges. Spread over more than two dozen groves in twelve counties of the inner north Coast Ranges, Sierra Nevada foothills, and Cascade Range, McNab Cypress regularly mingles with Chaparral and Foothill Woodland species like Gray Pine (*Pinus sabiniana*), formerly Digger Pine; Buckbrush (*Ceanothus cuneatus*); Knobcone Pine (*Pinus attenuata*) and manzanitas. Generally shrubby in form and versatile in soil tolerance, McNab Cypress can grow on clay, granite, basalt, gabbro, and a whole spectrum of rock types, including serpentine. While all California cypresses share similar roundish woody cones and have foliage that looks much alike—green overlapping scales, tightly fitted like shingles on a roof—McNab's branchlets usually differ from the rest in forming flatter sprays.

Wherever Sargent Cypress grows, California's state rock, serpentine, usually lies beneath it. Virtually restricted to soils derived from the silky-sheened green rock, of which California has the most abundant supply in North America, Sargent Cypress appears intermittently the length of the Coast Ranges.

Some of its finest stands occur at "The Cedars," on Austin Creek in Sonoma County, where thousands of compact, shrubby trees fill a massive serpentine outcrop, and on Cuesta Ridge West, north of San Luis

Obispo, where an almost pure, even-aged forest extends for more than 2 miles (3.2 km).

Sargent Cypress associates with many kinds of communities in its long range, frequently with Chaparral and Foothill Woodland, where it mixes with Gray Pine, Leather Oak (*Quercus durata*), various manzanitas and *Ceanothus*, and Coulter Pine (*Pinus coulteri*). It can be found on rocky outcroppings within the summer fog belt of the Redwood Forest, as well as among mixed hardwoods.

Serpentine is widespread in the state, and the plants that can tolerate its poisonous effects find lessened competition on its slopes. Soil derived from it tests drastically low in calcium and potassium, elements essential for plant growth, and tests potentially toxic in magnesium, nickel, chromium, cobalt, and sometimes asbestos.

All these features merge into a formidably inhospitable medium that only certain specialized plants can handle. Those that manage, like Sargent Cypress, are somehow able to extract enough calcium and other essential elements from the deficient soil to eke out an existence. Sargent Cypress has been surviving successfully long enough to have a butterfly that specializes on it. The delicately brown Muir's Hairstreak (*Mitoura nelsoni muiri*) lays its eggs only on Sargent, and its larvae feed on that cypress alone.

Two of the most restricted cypresses in the state grow in San Diego County, the Cuyamaca Cypress (*Cupressus arizonica* ssp. *arizonica*) and the Tecate Cypress (*Cupressus forbesii*). Cuyamaca wins on scarcity with only a single known population among dense Chaparral and sprinkled Coulter Pines at the 4,000-foot level (1,200 m) in the upper King Creek region of Cuyamaca Peak.

Tecate Cypress is more widespread, growing among Chaparral at 2,000 to 2,500-foot elevations (600 m) on Otay Mountain, as well as on a number of other north-facing Chaparral slopes south into Baja California. The areas where it lives usually hold more soil moisture than their surroundings and have suffered less frequent fires.

Tecate's clustered cones often remain closed on the tree for 10 years or more until fire vaporizes the resin and chars the cone scales. The scales separate after the fire has passed, gradually releasing the unburned seeds, roughly 100 per cone, over 4 to 5 months.

A number of rare and endangered plants share the various Tecate Cypress stands, as do Sage Sparrows (*Amphispiza belli*), the increasingly scarce San Diego Horned Lizard (*Phrynosoma coronatum blainvillei*), Mountain Lions, and Golden Eagles (*Aquila chrysaetos*). The Nature

Conservancy secured several southern California Tecate Cypress stands in the early 1990s. One of them, the northernmost site in Coal Canyon, Orange County, contains the largest Tecate Cypress on record, 150 years old and 35 feet high.

The most northerly cypress in the state, Baker Cypress (*Cupressus bakeri*), ranges from Plumas County to beyond the Oregon border. It is named for Milo Baker, Santa Rosa Junior College botanist, who discovered it growing on Modoc Plateau lava in 1898. This species reaches the highest elevation and the coldest climate of any closed-cone cypress within the state, over 7,000 feet (2,100 m), in Red Fir Forest north of Wheeler Peak. Other groves grow within Mixed Conifer Forest at 5,000 feet (1,500 m) in Shasta County, as well as in the Siskiyou Mountains, sometimes associating with Juniper Woodland and Sagebrush Scrub.

Attractive trees with straight central trunks, dark red, smooth branches, and foliage pleasantly aromatic when crushed, they may reach heights of 70 feet (21 m) and diameters of over 4 feet (1.2 m).

In the southern Sierra Nevada's Kern River drainage live less than a dozen groves of Piute Cypress (*Cupressus arizonica* ssp. *nevadensis*) Small trees, they intersperse with arid Foothill Woodland and Chaparral, and occasionally with pinyon pines and junipers. Capable of using a variety of soils, they reach their best growth between 4,000 and 6,000 feet (1200 to 1800 m) on Bald Eagle Peak south of Bodfish.

The Santa Cruz Mountains hold four populations of another rare cypress, Santa Cruz Cypress (*Cupressus abramsiana*). A survivor from glacial times, this endangered species lost its largest specimen in 1983 when vandals cut down a tree rated Champion on the American Forestry Association's National Register of big trees. The champion stood 75 feet tall (22 m), with limbs spreading 60 feet (18 m) and a trunk more than 5 feet (1.7 m) in diameter.

To save the Santa Cruz Cypress and its habitat, the California Nature Conservancy and the state Wildlife Conservation Board in 1989 secured protection of over 500 acres in the Bonny Doon area. Ecologically rich, this reserve harbors an abundance of rarities, including Peregrine Falcons (*Falco peregrinus*), Golden Eagles, Purple Martins (*Progne subis*), an extremely primitive solitary bee (*Colletes kincaidii*), and other wildlife of special concern.

In addition to Santa Cruz Cypress, the reserve contains Redwood river forest, willow thickets, mixed Chaparral, and Ponderosa Pine forest (*Pinus ponderosa*) far from its more common home in the Sierra

Nevada. The sandstone soil laden with fossilized sand dollars is unusual and supports some plants found nowhere else in the world.

Undramatic as a group, cypresses generally occupy poor, rocky soils in many unnoticed parts of the state and are commonly called "cedars" in place names. Port Orford Cedar and Alaska Cedar, also known as Lawson Cypress and Yellow Cypress respectively, add a slightly different dimension to the cypress group. Formerly considered false cypresses (*Chamaecyparis*), they have been reclassified in *The Jepson Manual* as true cypresses (*Cupressus*). Both favor more moist areas of northwestern California forests and are discussed in those habitats.

Together, all of the state's cypresses give California a sizable representation among the 15 kinds of cypresses found in North America.

Sharing many of the same inhospitable habitats as the cypresses, Knobcone Pine (*Pinus attenuata*) outlasts all of them in the length of time its cones stay closed. The whorls of knobby, asymmetrical cones that encircle Knobcone Pine trunks and major branches become entombed as the tree grows, almost never opening until a fire sweeps through the grove. Then, even while the trees are still smoking, the seared cones burst open like machine-gun fire and shoot thousands of pine seeds onto the charred earth.

As fire wipes out their shrubby competition, it opens the floor to sunlight and drops the necessary mineral ash seedbed to start a new, even-aged stand that replaces the blackened parents. So partial are Knobcones to fire as a rejuvenator that if they don't get one within a 50 year span, they begin to die and rarely last to 75 years. If squirrels cut open their cones, the thinly coated seeds that drop out die quickly without the sterilized mineral ash bed.

The most widespread of California's closed-cone pines, Knobcones range sporadically nearly the length of the state on steep slopes, dry ridges, and thin, impoverished soils—substrates that most other pines can't tolerate. The largest stands grow primarily on serpentine in the northwest, where they intermix with many forest types. But they can also handle sandy and granitic soils, and, in the Santa Ana Mountains and other outer Coast Range locations, utilize seasonal fog drip. Commonly occurring with Chamise (*Adenostoma fasciculatum*), manzanitas, and oaks in hot, dry foothills, they frequently occupy the transition area between lower Chaparral-Woodland and higher coniferous forests in the Sierra Nevada, Coast Ranges, Cascades, and San Bernardino Mountains.

Figure 19. Knobcone Pine's stem-encircling cones open only
           with heat from direct sun or after fire sweeps through
           their groves.

Admired more for their toughness and adaptability to fire than for
their beauty, Knobcone Pines usually grow as scraggly trees with thin,
faded three-needled foliage that offers scant shade. In one locale in the
Santa Cruz Mountains, near Año Nuevo–Swanton, where their range
overlaps that of Monterey Pine, the two hybridize, concrete evidence
of the close relationship that appears to exist among all the closed-
cone pines.

California owns no monopoly on closed-cone pines. The Rocky
Mountains have their fire-type Lodgepole Pine, the upper Midwest and
Canada their Jack Pines (*Pinus banksiana*), New Jersey its Pitch Pine
(*Pinus rigida*), and Florida a Sand Pine (*Pinus clausa*). But California's
four pines and ten cypresses far exceed any other area of North America
in species of closed-cone conifers.

Torrey Pines (*Pinus torreyana*) do not fit the closed-cone pine category
in the strictest sense, since their broad cones turn chocolate brown and
open after three seasons. But they drop only part of their seeds on
ripening and retain some within the heavy cones on the trees for up to

15 years, seeds that when released from fire-killed trees, successfully germinate.

The rarest pines in the world in numbers of individuals, Torreys grow natively today in only two very limited places—Torrey Pines State Reserve near San Diego and 175 miles northwest (280 km) on Santa Rosa Island. Small trees, 20 to 30 feet high (6 to 9 m), they bend and twist out of the ancient sandstone cliffs above the Pacific Ocean at the Reserve in picturesque open forests. The Coastal Sage-scrub community in which they live is a tough one, arid, wind-blasted, easily eroded, and fragile.

Ocean fogs bring some relief, condensing water droplets on the 8 to 12-inch-long needles (20 to 30 cm), among the world's longest and strongest. The trees survive life spans of about 100 years under these stressful conditions. Their Chaparral-like understory includes shrubs such as Wart-stemmed Ceanothus (*Ceanothus verrucosus*), Black Sage (*Salvia melifera*), and Lemonade Berry (*Rhus integrifolia*), along with Chaparral animal life.

Like many of California's choice wild places, Torrey Pines State Reserve owes its existence to dedicated conservationists who recognized the uniqueness of the trees on the coastal bluffs and organized campaigns to preserve them. Charles Parry, the explorer-botanist who discovered the San Diego site in 1850 and named the trees for his colleague and teacher John Torrey, returned to the area 33 years later to urge the city to save this special forest from the logging, grazing, and general despoliation that was going on. Philanthropist Ellen Browning Scripps, naturalist Guy Fleming, and local groups carried on the battle from one generation to the next. Eventually, hundreds of bake sales, craft fairs and carwashes later, they raised enough private contributions to match state park bond funds.

Today, Torrey Pines State Reserve and adjacent protected areas form an unduplicatable unit of the California State Park System. The Parry Grove and Guy Fleming Trails lead through a succession of wild flowers from February to July, under the sprawling Torrey Pines.

Recent studies on these pines revealed something very unusual in the species. Conifers normally show tremendous genetic variation between individuals. In the western United States only Western Red Cedar is known to have a variability as low as four percent.

Torrey Pine has a genetic variability of zero. The fifty-nine genes of the mainland trees are identical. Internally, the trees are as uniform as clones; externally, the environment shapes their form. Somewhere in the Torrey Pine's past, the genes that gave it diversity were wiped out. It survives with genes that allow no leeway for future major environmental changes.

The same situation exists in the Torrey Pine forest on Santa Rosa Island. All the trees there (*Pinus torreyana* ssp. *insularis*) are as identical as clones with each other. But the entire group differs in two of its fifty-nine genes from the mainland trees. Getting more rain and fog, the island Torrey Pines grow larger, with greater spread of branches and more massive cones.

6

Foothill Woodland

The roads that head east and west out of California's Great Central Valley gradually leave cities and flat farmland behind and climb into low rolling hills where round-headed oaks scatter themselves among the grasses of an open parkland. As the hills grow higher, the foothill's silvery grayish pines spread lacy crowns above the oaks, shrubby slopes of Chaparral, and the tangled vegetation of creeks and ravines. This pine-oak Foothill Woodland, intermingling with Chaparral, stretches green and golden all the way up to the darker Mixed Conifer Forest of the next higher mountain belt.

Uniquely Californian communities, Foothill Woodland and Chaparral thrive in a Mediterranean climate of mild, wet winters and hot, dry summers. From May to October not a drop of rain usually splatters the thirsty earth. Every plant and animal here must successfully conserve the water it needs through five summer months of baking sunshine with temperatures hovering near 100°F (38°C). The plants adapt with needles and tough, waxy leaves that reduce evaporation. Larger leaves carry sunken pores, while roots reach far down for water.

Gray or Foothill Pine (*Pinus sabiniana*), formerly known as Digger Pine, is the dominant conifer of Foothill Woodland, and the only one in most places. It survives by being deep-rooted and drought resistant, able to persist on as little as 10 inches (25 cm) of rain. In summer when heat radiates off the tombstone boulders and parched grasses lie dormant under the drooping needles of this loosely foliaged tree, foothill transients might well wish for the cool shade of the darker, tighter fo-

Map 6. Foothill Woodland.

liaged pines at higher elevations. But in winter and spring, when the higher conifers are immersed in cold and snow, Gray Pines spread their veil-like, open crowns over the greening grasses and early wildflowers of the rolling hills with a grace that fits the landscape.

Shaped very differently from most pines, Gray Pines usually divide into many trunks a short distance above the ground. Some botanists have compared their form to a loose broom or a lyre. John Muir commented that they branched more like an elm. Their color blends with the general silvery green tones of other hill vegetation. The supple needles, 8 to 12 inches long (20 to 30 cm), are pale gray-green, borne in clusters of three.

The large, heavy cones, shaped a little like a pineapple, weigh up to four pounds. When mature, the lower cone scales end in stout, curved, downward-pointing spikes covered with pitch. They are a menace to handle. But the prizes they contain, tasty pine nuts nestled in pairs at

Figure 20. Gray Pine (*Pinus sabiniana*). Heavy, spiky cones grow
amidst the sparse silvery foliage of this Foothill
Woodland conifer.

the base of each husky cone scale, are nearly twice the size of the better-
known pinyon pine nuts. Western Gray Squirrels regularly gnaw off the
closely set scales to get at the sweet kernels.

California Indians of the foothill country relished the nuts' flavor.
John Muir described how, at harvest time, the men climbed the pines
and knocked off the bulky cones with sticks, not an easy task, as the
cones hung on tenaciously. Women gathered the resiny cones and
roasted them until the scales opened to expose the hard-shelled nutritious seeds.

They utilized other parts of the tree as well. Cutting out the soft
inner core of the green, first-year cone, they devoured it raw. In spring,

inner bark and young buds of the pine were choice food items, and they occasionally chewed the gummy resin exuding from tree wounds.

The Indians of the western Sierran foothills were primarily of the Miwok and Yokuts tribes. Extremely adept at living off the land, they knew when and where to harvest the seeds of different kinds of grasses and herbs, which plants were poisonous, which roots and corms were safe and tasty and when they were ready for digging. Because the Miwok and Yokuts dug for part of their diet, some of the early Gold Rush pioneers in California dubbed them "Diggers" and called the foothill tree from which they got their pine nuts "Digger Pine." Both terms were often used in derogatory fashion, with no understanding of the knowledge required to dig successfully for your dinner and no appreciation of the Miwok and Yokuts cultures which passed on food survival wisdom by memory from generation to generation.

The Miwok called their pine-nut tree "Ghost Pine" for the way it emerged, tall and silvery white, from a hillside of Chaparral with the sun behind it. The tree has also been called Bull Pine locally and Gray Pine for its predominant color. Digger Pine remained the established name in botanical books for decades.

Many of today's Native Americans find the "digger" association offensive and prefer an alternative name for the tree. Hence, the old names, Gray and Digger Pine, used in George Sudworth's *Forest Trees of the Pacific Slope* in 1908, have been changed to a choice of Gray or Foothill Pine in *The Jepson Manual: Higher Plants of California, 1993*.

By any name, the airy conifer plays a major role in foothill life. Growing in thin, scattered stands, it almost never forms closed forests as many other conifers do. Whether on canyon walls or on more gentle grassy hills, it occurs most commonly between elevations of 1,000 to 3,000 feet (300 to 900 m) in both the Coast Ranges and the western Sierra Nevada, often on soil derived from serpentine rock. Southern Monterey County is the only region where it comes close to the coast.

In canyons of the Merced, Kern, and Trinity rivers, it leans precipitously over the steep slopes, often at right angles to the hillsides. Plant and animal life is rich in these foothill canyons. The river banks support White Alders (*Alnus rhombifolia*), Big-Leaf Maples, Western Sycamores (*Platanus racemosa*), willows and poplars, along with shrubs and wild flowers in season. Redbud (*Cercis occidentalis*) adds reddish purple flowers in spring and Button Bush (*Cephalanthus occidentalis*) balls of white blossoms in early summer.

The plant cover draws many birds, including nesting Warbling Vireos (*Vireo gilvus*), Yellow Warblers (*Dendroica petechia*), and Black-headed Grosbeaks (*Pheucticus melanocephalus*). Dippers fly low over the rushing streams and build nests behind waterfalls. Belted Kingfishers raise broods in the banks. In summer the liquid notes of the Canyon Wren (*Catherpes mexicanus*) cascade down the canyon walls.

At night the Ringtail leaves its cave up in the cliffs and descends the canyon to drink and hunt for Deer Mice, which are sniffing out their own prey—grasshoppers, caterpillars, beetles.

It is still a puzzle why Gray Pines are missing from a 55-mile gap between the South Fork of the Tule River and the Kings River in the southern Sierra Nevada. Their usual companion tree over foothill range, Blue Oak, is there, as are California Buckeyes (*Aesculus californica*), Interior Live Oaks and regular Chaparral associates. No soil differences are apparent. Botanists have been baffled by this mystery since Josiah Whitney mentioned it in 1865.

With pines or without, much Foothill Woodland is a land of oaks, with two distinctively Californian types especially notable—Valley Oak and Blue Oak. Valley Oaks (*Quercus lobata*), once common in deep, fertile soils of the Central Valley, forming luxuriant groves along segments of its rivers, are now much less visible there. But they have always climbed into the foothills, seeking the more moist, deeper soils of both the Coast Ranges and the western Sierra Nevada. They form open savannas with Blue Oaks in grassland habitats and join the evergreen live oaks and pines of both ranges higher up. Valley Oaks may grow with Coulter Pines in the south Coast Ranges at elevations up to 5,000 feet, with Jeffrey Pines in the Tehachapis, and with other mixed species as far south as the San Fernando Valley. Wherever they occur, their stately forms lend a parklike appearance to the landscape.

But it is the Blue Oak (*Quercus douglasii*), whose range overlaps that of Gray Pine almost totally, that shares dominance with its conifer companion in much of Foothill Woodland. Both trees prosper on the driest hillsides, utilizing root systems up to 30 feet (9 m) deep. Both spread their branches widely and grow to modest heights compared to most pines and oaks. Both produce foliage of similar pale tones, the pine's needles grayish green, the Blue Oak's small, oblong leaves bluish green. And both look their most luxuriant in spring, at the end of the rainy season when grasses and wildflowers blanket the hills.

The grasses are mostly introduced annuals, wild oats (*Avena* spp.), Foxtail Brome (*Bromus rubens*), Ripgut Brome (*Bromus diandrus*), and others, which invaded the hills during and after the Mission Period, 1769–1824. As the Spanish Fathers set up gardens around their string of California missions, they unloosed seeds brought from their Mediterranean soils. Ballast from ships added more.

Many of the seeds were of tough annual grasses adapted to the aridity and the heavy, constant grazing and trampling of domestic livestock in their homeland. When the cattle boom of the mid-1800s combined with years of extreme drought in California, the alien annuals replaced the native perennial grasses in droves. Today's foothill grasses are a cosmopolitan world collection.

Native foothill wildflowers fared better than native grasses. Spring still spawns yellow waves of goldfields (*Lasthenia* spp.), *Blennosperma nanum*, Butter-and-eggs (*Orthocarpus erianthus*) and Johnny-jump-ups (*Viola douglasii*) around and under the Blue Oaks. Tidy-tips (*Layia platyglossa*) still wave among fields of white popcornflowers (*Plagiobothrys* spp.), blue brodiaeas (*Brodiaea* spp.), and fiery California Poppies (*Eschscholzia californica*).

The spaces under the Blue Oaks often house a much richer grass and forb assortment than the open areas. Being deciduous, Blue Oaks drop their leaves each autumn. These, along with the tree's numerous fallen, brittle twigs, form a litter which decomposes into a soil higher in nutrients, organic matter, and water-holding capacity than the surrounding grassland.

Grasses and forbs under the canopy stay green much longer in spring because of the shade and remain protected from the extreme heat of summer. Cattle habitually chew their cuds midday under the Blue Oaks, favoring the shade and finding a forage high in nutritional content well after it has diminished in the baked open areas. However, at times, Blue Oak roots form dense mats under the crown and restrict the heavy grass cover.

The evergreen live oaks of the Foothill Woodland offer a deeper, wider circle of shade the year around. Their dark green, shiny, leathery leaves separate them readily at a distance from the dull blue-green leaf cover of the Blue Oaks. And their dome-shaped crowns, with branches right to the ground where unbrowsed by cattle or deer, are much broader than tall.

Coast Live Oaks grow primarily in valleys and foothills of the Coast Ranges, although they extend into the Central Valley in a few places. In Mendocino County they hybridize with Interior Live Oaks, which spread into both north and south Coast Ranges, as well as into the broad foothill belt along the western slope of the Sierra Nevada.

The most widely distributed oak in the state, Canyon Live Oak, sometimes called Maul Oak, lives on steep, rocky canyon slopes in many forest habitats, often side by side with Interior Live Oaks. Where their ranges overlap, the Canyon Live Oak can be distinguished from the Coast Live Oak and Interior Live Oak by the grayish undersides of its year-old leaves and its fuzzy golden acorn cup. The Coast Live Oak has leaves that are usually convex on the upper surface and, on the lower surface, contain patches of fuzz in the angles where the midrib joins larger side veins. Interior Live Oak leaves are usually flattish and light green below, with no fuzz. Leaf margins vary from spiny to smooth on all of them.

Many oaks hybridize. Blue Oak sometimes hybridizes with Oregon Oak, which replaces it at the northern end of its range. At its southern end, below the Tehachapis, a coniferless southern oak woodland dominated by Engelmann Oaks (*Quercus engelmannii*) and Coast Live Oaks takes over.

Plant communities almost never follow precise lines of distribution. Topography, soil, local climatic conditions, fire, and other factors inevitably allow for variances and transitions. In the Tehachapi and Piute ranges, Blue Oak and Gray Pine mix with California Junipers (*Juniperus californica*), Singleleaf Pinyon Pine (*Pinus monophylla*), and sometimes Big Sagebrush (*Artemesia tridentata*) in a transition phase to the Great Basin Province of the eastern Sierra. In the Pit River area of northern California, east of the Cascades, isolated Blue Oak stands grow amid Northern Juniper Woodland. Near the rocky summit of Mt. Diablo in central California, junipers provide a shady campground adjacent to Blue Oaks.

The wildlife of Foothill Woodland is often richest in the woods where the pines and oaks intermingle with California Buckeyes, a shrubby undergrowth of Buckbrush, California Coffeeberry (*Rhamnus californica*), Toyon or Christmas Berry (*Heteromeles arbutifolia*), and tangling vines of Wild Grape (*Vitis californica*), Pipestems or Virgin's-Bower (*Clematis lasiantha*), and Western Poison Oak (*Toxicodendron diversilobum*).

Buckeyes stand out in the woodland at any season of the year. First to cover themselves with fresh green palmate leaves in late winter, as early as February, they later produce showy candlelike white flower

Figure 21. Where Foothill Woodland and Chaparral intermix,
Gray Pines (top) overlook deciduous California
Buckeye (left) and dense Chamise (right). The Wrentit (center left), Gray Fox, and Brush Rabbit find
plenty of cover.

spires in May. In many cases only one flower from each white spire eventually produces a fruit. And while this single fruit is forming over the hot summer, buckeyes' long leaves turn brown and dry up, their own special adaptation to conserving water within the stem and not losing it through leaf transpiration.

By autumn, the small, sprawling trees stand bare, their crooked silvery branches dangling pearlike fruits. When the fruits' tough husks crack open, each reveals a smooth, glossy, rich brown seed up to 2 inches (5 cm) across, as captivating aesthetically as the eye of a deer— a "buck eye."

While they are visually appealing, the seeds are poisonous when raw. California Indians formerly used them to capture fish. By grinding the seeds and emptying them into pools, they stupified the fish, which floated to the surface and were easily picked up. Pouring water through the ground nuts leached out the poison before the nutritious mixture was cooked.

Little grows under a buckeye. But in the dense thickets surrounding it, in many foothill areas, lives a plant that is equally toxic in its own way—Poison Oak. One of the most varied and widespread plants in California, Poison Oak grows as a shrub, a tree, or a vine in both shady forests and open sunny places. Attractive at any season, its new spring leaves emerge shiny and bronze, turn green in summer and scarlet in autumn before they fall. Nearly always in groups of three, the leaves furnish nutritious forage for deer, cattle, and horses. But many people touch them at their peril.

All parts of Poison Oak, except possibly the pollen, contain the clear, heavy oil, urushiol, a force to be reckoned with. So incredibly potent is the oil that a pinhead amount of it can cause rashes, blisters, and itching in sensitive people. So long-lasting is the oil that botanists handling century-old samples have developed serious dematitis. Fire fighters who breathe smoke from flaming Poison Oak can suffer dangerous lung infections, head-to-toe dermatitis, even death from a swollen throat. People vary enormously in their sensitivity and can lose an early immunity later in life.

While urushiol is confined to canals inside the plant's leaves, stems, and roots, it oozes out readily when these are bruised or chewed by insects or brushed against by a dog or a pair of Levis or boots. All of these may transmit the oil for weeks or months afterwards. Urushiol takes only about ten minutes to penetrate the skin, so a quick washing or absorbing the oil with dirt constitute the simplest emergency field measures.

California Indians apparently enjoyed an immunity to the oil as they wove baskets from the supple stems of Poison Oak. The fresh juice produced an excellent black dye and was also used as a cure for ringworm. The leaves served as wrappings for acorn meal during baking.

Poison Oak's white berries, rich in vegetable fat, form a favorite food of nearly forty kinds of California birds, including Band-tailed Pigeons, Mourning Doves (*Zenaida macroura*), and crows. The seeds pass through the birds' digestive tracts unharmed to spread the plant wherever they land. Among their unwitting distributors in Foothill Woodland is a woodpecker with black and white cross bars on its back which resemble a ladder—Nuttall's Woodpecker (*Picoides nuttallii*).

There are other "ladder-backs" in the southeastern and southwestern United States, but Nuttall's Woodpecker largely restricts itself to California woodlands and adjacent riparian environments. Here it forages chiefly on Blue and Live Oaks, gleaning for surface and subsurface insects—wood borers, leaf beetles, weevils, and true bugs. Instead of hitching up the trunk, using its tail as a prop as most woodpeckers do, Nuttall's more frequently circles branches, creeping in a humped posture, probing and pushing off bark scales. It hops into leaf and twig clusters, scanning them for insects, perches crosswise on small limbs with its tail in the air as a balancer, or hangs upside down gathering berries from Toyon, Blue Elderberry (*Sambucus mexicana*), or Poison Oak.

The bird's namesake, Thomas Nuttall, an Englishman, came to America in 1808 at the age of twenty-two, enthralled, as he said, with "The Goddess Flora." Without even finding lodging on the day of his arrival in Philadelphia, he set out to do what he came to do—naturalize. As Bil Gilbert documents his story, Nuttall spent the next 35 years exploring America's wilderness frontiers, with expeditions or on his own. The country probably never saw a more inept field botanist. Racked by malaria much of the time, unable to steer a canoe effectually, or to swim, incapable of riding a horse competently, or setting up a safe, comfortable camp, or shooting a gun, he nevertheless survived swamps, Indians, starvation, and getting lost as he single-mindedly pursued his passion for natural history.

Somehow luck and his companions looked after him, and by his mid-forties he had published the most comprehensive reference book of the times on American plants and birds. He later filled the chair of

Botany and Ornithology at Harvard and had his name immortalized in Nuttall's Cottontail, Nuttall's Poorwill, Nuttall's Dogwood, as well as Nuttall's Woodpecker.

The Nuttall's Woodpeckers of California woodlands sometimes have to compete for nesting holes with Ash-throated Flycatchers (*Myiarchus cinerascens*), Plain Titmice (*Parus inornatus*), Hairy Woodpeckers (*Picoides villosus*), and House Wrens (*Troglodytes aedon*). But they leave the oaks' acorn crop to the larger, noisier Acorn Woodpecker (*Melanerpes formicivorus*). Only at "sap trees" do the two species sometimes clash with each other and other birds over the sweet fluids oozing from sapsucker holes.

Acorns are a favorite food of countless foothill birds and mammals, but the Acorn Woodpecker specializes in them. One of the truly unusual birds of the world, this garrulous resident of oak and pine-oak woodlands of southwestern United States, Mexico, Central America, and California is a notable exception to the way most birds live, mate, breed, and feed.

Instead of a pair raising a family, feeding and guarding them as a self-contained social unit and dispersing after the young leave, Acorn Woodpeckers in California follow an entirely different life style. They live in territorial groups of up to a dozen or more year round. Their territory centers around a granary of acorns and other nuts, known as mast, stored in holes they drill in dead or partially dead trees. The storage holes are as important as the acorns.

All winter they live on the stored mast and defend it from squirrels, jays, and other intruders. Cooperatively drilling new holes for future expansion, they also move acorns into smaller holes as the drying acorns shrink.

Over a century ago, a mistaken theory about these Acorn Woodpeckers emerged. Called the "grub theory," it surfaced in an 1866 article by C. T. Jackson in the *Proceedings of the Boston Society of Natural History*. He claimed that these birds deliberately store acorns infested with young insect larvae so that they can return later and eat the mature fat grubs.

Repeated over the decades in various publications, the theory was refuted forcefully by field observations of ornithologists C. Bendire and Lyman Belding in 1895 and 1904 and was finally put to rest scientifically in 1911 by Beal's analysis of the stomach contents of eighty-four Acorn Woodpeckers. He found abundant acorns, with larvae almost entirely wanting. Recent studies indicate that the birds reject any acorns

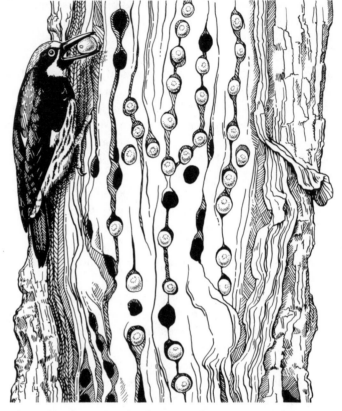

Figure 22. Acorn Woodpeckers store their communal winter
acorn supply in granary trees. They fit the acorns into
the holes like fingers into a tight glove.

that have been invaded by insects. Such an enticing story, however, does not die easily and still frequently pops up in local folklore.

Other facets of the birds' lives are equally remarkable. When spring arrives, they also breed cooperatively. The sexually mature birds in each group divide into breeders and nonbreeding "helpers." The breeders commonly share mates, and females all lay their eggs in one active nest cavity. The young Acorn Woodpeckers, true group progeny, are defended and fed insects by the entire group.

As the young mature, they usually become next year's helpers and stay on in the territory. When summer's abundance of flower nectar, catkins, tree sap, and insects wanes and the green acorns of early autumn have turned brown, the first-year birds' cupboards would be bare for the winter if they were not part of a group that owns a granary.

Some granaries hold more than 30,000 holes. They are usually in the largest dead or dying pines, with soft bark for easy drilling, but the birds use oaks, telephone poles, fence posts, cabins, or whatever wood is available, when pines are scarce.

In southeastern Arizona, where the annual acorn crop is sparse and variable, Acorn Woodpeckers migrate to Mexico when they run out of winter mast.

Although Acorn Woodpeckers belong to a distinct minority in the bird world in cooperative breeding, they are not the only birds to survive this way. Others include the Florida Scrub Jay (*Aphelocoma coerulescens*), the Green Woodhoopoe of Africa (*Phoeniculus purpureus*), the Mexican Jay (*Aphelocoma ultramarina*), some African bee-eaters and South American wrens.

The list of birds and mammals that thrive on acorns, along with the insects, nuts, and berries, of California's Foothill Woodland is a long one. Over 35 mammals and nearly 100 kinds of birds include acorns in their diet. Among the better-known species, Mule Deer and Western Gray Squirrels rely so heavily on acorns that a poor annual crop can limit their populations. Band-tailed Pigeons, Scrub Jays and introduced Wild Turkeys (*Meleagris gallopavo*), among others, feed extensively on acorns. Near Ash Mountain in the southern Sierra, David Graber happened onto a rare sight in 1986—dozens of Black Bears sitting in Blue Oak trees, eating acorns.

Robins, Western Bluebirds (*Sialia mexicana*), Phainopeplas (*Phainopepla nitens*), and Cedar Waxwings (*Bombycilla cedrorum*) are especially fond of the sticky berries of the Oak Mistletoe (*Phoradendron villosum*). Mistletoe seeds germinate after passing through the birds' digestive tracts and are spread by them from tree to tree. Oak Mistletoe taps the xylem line of its host oak for water and any minerals in the water, but it is a green plant that makes its own food.

Gray Pine occasionally harbors a different kind of mistletoe, the yellowish Dwarf Mistletoe (*Arceuthobium campylopodum*), a true parasite that steals its total nourishment from the pine.

The leaves, twigs, and bark of woodland trees provide rich insect-catching grounds for Bushtits, Hutton's Vireos (*Vireo huttoni*), and Northern Orioles (*Icterus galbula*). Abandoned woodpecker cavities house Plain Titmice, White-breasted Nuthatches (*Sitta carolinensis*), and American Kestrels (*Falco sparverius*). Great Horned Owls sometimes take over last year's stick-nest platforms of Red-tailed Hawks (*Buteo jamaicensis*) and occasionally nest in Gray Pines amidst a rookery of squawking Great Blue Herons (*Ardea herodias*).

Gray Fox den in larger tree cavities. Mountain Lions stalk deer through the hills in the fall and winter. Coyotes and Bobcats hunt cottontails, Black-tailed Jackrabbits (*Lepus californicus*), and California Ground Squirrels (*Spermophilus beecheyi*) in the grassy areas. Ground squirrels sometimes defend their young from Pacific Gopher Snakes (*Pituophis melanoleucus catenifir*) and the Western Rattlesnake (*Crotalus viridis*) by kicking dirt into the snake's face.

Rattlers, in turn, fall prey to the black-and-white-banded Common King Snake (*Lampropeltis getulus*), which is immune to their poison, and to the Racer (*Coluber constrictor*). The slim, speedy racer, hunting with head held high, especially relishes lizards. It seizes blue-bellied fence lizards and long-tailed Southern Alligator Lizards (*Elgaria multicarinatus*) with its sharp, recurved teeth and swallows them whole. Both lizards often hide out in the bulky stick-nest mounds of wood rats, joining salamanders, *Triatoma* bugs, and other refugees.

From such dense brush, whether in woodland or Chaparral, the staccato song of the Wrentit (*Chamaea fasciata*) rings out much of the year. Small brownish birds with long tails, Wrentits are easy to hear but hard to see. Unusual among songbirds, they mate for life and are virtually inseparable. They often sleep together on a branch, blending into a single ball of feathers.

All these interrelations worked out their own balance for millions of years in natural Foothill Woodland. But in the last century things have drastically changed. The vast number of sheep and cattle let loose in the range country in the late 1800s nibbled the grasses to their roots, devoured acorns and seedlings of the oaks, and trampled the few seedlings that escaped. Their heavy hoofs compacted the ground, reducing its ability to absorb water. Reproduction in Blue Oaks and Valley Oaks virtually ground to a halt.

The oaks' problems were compounded by the exotic annual grasses introduced during the Mission Period. Generating enormous yearly seed crops, the introduced annuals far outproduced the remaining native perennials and drew seed-eating mammals such as ground squirrels and field mice to the range. All, including Botta's Pocket Gopher (*Thomomys bottae*), ate acorns and oak seedlings.

In addition, the introduced annuals proved to be tougher competitors for water and nutrients than the perennial grasses had been. Oak seedlings on grazing lands today stand roughly a zero chance of survival. The only young trees coming up are in scattered roadside or ungrazed areas. Blue Oaks and Valley Oaks are represented almost entirely by old-timers.

Mature Blue Oaks face another serious threat. Many have been cut down by ranchers under the impression that this will yield more forage for their livestock. Research actually shows the opposite to be true. The Blue Oak is a very beneficial tree. Don Duncan of the San Joaquin Experimental Range collected data from thousands of sample plots to show that from 15 to 100 percent more forage is produced under Blue Oaks than in open areas. Of equal importance, the quality and nutritional value rank higher, as does the stored energy. Vegetation under Blue Oaks also differs floristically from its surrounding grassland, so that certain plant species are more abundant under the canopy and others in the open, variations noticed favorably by both ecologists and cattle. James Griffin and V. L. Holland have verified the diversity provided by Blue Oaks over decades of field studies.

The oaks face destruction from axe, chainsaw, and bulldozer in ever-growing volume. The foothills are one of the most rapidly developing areas in California. Houses and golf courses spring up in one subdivision after another on rolling land once dotted with oaks, pines, and tangles of Wild Grape and bisected by creeks and ravines. Wood stoves demand fuel; the oaks that crash to earth to provide it become part of a past vanishing all too quickly.

Unless action is taken, Foothill Woodland, like the once endless prairies, will not roll scenically on, mile after mile, forever. The least disturbed segments of it could still be preserved in state and county parks and subdivision greenbelts. Large landowners and ranchers with a long-term view may be willing to create pine-oak preserves on private lands, under conditions similar to the Williamson Act on agricultural land. Ecologically minded citizen groups could be recruited to set up replanting programs, help the seedlings through the first vulnerable years, and restore them to their historic range. This is happening with Valley Oaks in The Nature Conservancy's Cosumnes River Preserve.

State legislation passed in 1989 requests all state agencies with land use authority to assess the effects of their land decisions or actions on oak woodlands and to adopt policies to preserve and protect oaks under their jurisdiction. These agencies will require monitoring by alert. knowledgeable watchdogs, such as members of the California Native

Plant Society, the Sierra Club, Audubon societies, and the California Oak Foundation, established to preserve the state's oak heritage.

The Nature Conservancy has taken the lead on a number of prime oak sites, securing Engelmann Oak woodland on the Santa Rosa Plateau in San Diego County, Valley Oak woodland in the Kaweah Oaks Preserve near Visalia and in the Creighton Ranch near Corcoran.

Similar protective actions are needed to keep the pines and oaks of California's Foothill Woodlands a part of the state's natural diversity in the decades ahead.

7

Midmountain Forests
(Mixed Conifers)

California's midmountain forests boast a diversity of trees unequaled among the coniferous forests of the world. Many pine forests, such as those of the widespread Scots Pine in Eurasia, Benguet Pine in the Phillipines, Ponderosa Pine in Arizona, consist mainly of one coniferous species. California's forests, distributed over the middle elevations of the Sierra Nevada, southern California mountains, the Coast Ranges, the Cascades, and the Warner and Klamath Mountains, not only possess the greatest variety of coniferous trees but also some of the largest.

Known as midmontane or midmountain forests for their middle elevations, they are also called Mixed Conifer Forests for the generous mixtures of coniferous trees they contain. Since the species mix varies considerably from one region to another, we will use the Sierra Nevada Mixed Conifer Forest as an in-depth prototype of trees, birds, fire, and fungal ecology, and from that base explore the variations in other midmountain forests up and down the state.

Sierra Nevada Mixed Conifer Forest

The western slopes of the Sierra Nevada support some of the most magnificent Mixed Conifer Forests on the planet. The giants of the pine family live here. From the upper foothills to the Red

Map 7. Midmountain Forests (Mixed Conifers).

Fir belt, at elevations of roughly 2,500 to 6,000 feet (750 to 1800 m), Ponderosa Pines and Sugar Pines (*Pinus lambertiana*) raise their striking boles 200 feet or more (60 m) into the blue Sierran sky. With them grow equally impressive Incense Cedars (*Calocedrus decurrens*), White Firs (*Abies concolor*), and Douglas-Fir, lofting evergreen branches over the deciduous California Black Oaks and rich shrub and forb understory. Special groves hold venerable Giant Sequoia.

Ponderosa Pines, often called Yellow Pines, dominate these forests on dry, sunny sites, their mature yellow-orange bark divided into large plates separated by black cracks. Old-timers live for centuries and grow to ponderous size. A tree that John Muir discovered in Yosemite Valley reached 210 feet high (63 m) and 8 feet wide (2.4 m) before falling in the 1930s. A 600-year-old ponderosa near Cedar Grove in Kings Canyon National Park topped 200 feet (60 m) and 7 feet (2m) in diameter. But most are under 150 feet, and many are second-and third-growth specimens half that height, coming back from the heavy logging around the turn of the century.

Ponderosa Pine's 3 to 6-inch cones (7 to 15 cm), small enough to hold in the hand, prick with sharp, out-turned spines when fingers close around them. So variable is cone production that a mature tree may drop cones every year, every other year, or none at all, onto a forest floor already strewn with the pine's long, loose-lying needles.

Other conifers join the ponderosas in moist areas. Sugar Pine lends a special distinction to the midmountain forests. Tallest and largest of the world's 100 species of pines, it soars past the 200-foot mark with a style uniquely its own. Free of limbs for one-third of its great height, it extends long horizontal branches, intermingled with shorter ones, from the middle of the tree upward. The longest limbs are often near the crown, producing singular asymmetrical silhouettes (see frontispiece). Bob Fry measured one of these fallen limbs at 45 feet (13.5 m).

Short glistening needles, five to a bundle and 3 inches (7 cm) long, soften the boughs. On the ends of the outstretched branches dangle the world's longest pine cones—tan, brown, pitchy, beautiful pendants, 10 to 24 inches (25 to 60 cm) in length. From them, winged seeds float down in September of cone-bearing years, while the cones drop during high winds or storms of the following months. Only a few remain on the tree year round.

John Muir, who claimed Sugar Pine as an old friend, called it "the noblest pine" in the world's forests. He rated the sugar from which the tree gets its name "sweeter than maple sugar." The sugar exudes from wounds in the heartwood of trunks injured by forest fire or the axe. Candylike crystalline kernels form on the upper side of the wounds, white if fresh, brown if stained by a burn. Native Americans treasured them, eating them in small quantities, respecting their laxative effect.

The discovery and naming of the Sugar Pine by David Douglas rank as one of the legendary stories in botanical lore. The young Scot, one of the great exploring botanists of the early American West, had been intrigued during his Oregon expedition of 1826 by the large pine nuts carried by Indians in their food pouches. When they told him that the sweet nuts came from cones the size of breadloaves, he vowed to find the tree.

Drenched by rain, assaulted by winds that blew down his tent, suffering from fever, giddiness, and aches, he plodded through the untracked wilderness of the Willamette and Yamhill rivers, and finally, west of Roseburg, reached a grove of his sought-after pines, which indeed had immense cones hanging like loaves. On the ground nearby lay a massive blown-down Sugar Pine 215 feet long (65 m) and 18 feet thick (5.4 m).

The cones he coveted were high above, and the trunks lacked lower limbs for climbing. He took his gun and had shot three cones from the branches when eight Indians appeared.

His journal tells the story. "They were all painted with red earth, armed with bows, arrows, spears of bone, and flint knives, and seemed to me anything but friendly. I endeavored to explain to them what I wanted." But when one Indian strung his bow and another sharpened his flint knife, the Scot, without hesitation, stepped back six paces, cocked his gun, and pulled a pistol from his belt, "determined to fight for life" (Davies 1980:104).

For long minutes they faced each other in silence. At last, the Indian leader made a sign for tobacco, which Douglas indicated he would give them if they brought him some cones. As soon as the Indians left, he picked up his three cones and some twigs and beat a hasty retreat.

The mountains of southern Oregon, California, and northern Baja California are the only areas in the world where Sugar Pine grows. For a while, in the mid-1900s, it looked as though the tree might be endangered in its total range by White Pine Blister Rust (*Cronartium ribicola*). The rust, introduced into British Columbia in 1910 on a shipment of infected White Pine seedlings from France, moved from Washington and Oregon into California in less than 10 years.

Spread by fungal spores, the rust is windblown from infected five-needled White and Sugar Pines to gooseberry and currant shrubs (*Ribes* spp.). On the shrubs it develops through further stages, and spores are windblown back to the pines, where they kill pine needles, working down the branches 1 to 2 inches a year. If the rust reaches the main trunk and encircles it, the tree dies.

For several decades foresters eradicated gooseberries and currants, especially in Yosemite, to interrupt the cycle and to save the pines. In the 1960s, however, it seemed that the rust was not such a serious threat after all, reaching its southernmost station in Plumas County. Many Sugar Pines appeared to be genetically resistant. Most infections occurred far out on branches in the lower part of larger trees and became inactive before reaching the trunk.

That favorable prognosis was short-lived! Heavy outbreaks of blister rust in the Sierra in 1976 and again in 1983 caused deep alarm. Blodgett Forest's Sugar Pines in the northern Sierra near Georgetown, relatively uninfected prior to the 1976 attack, became 80 percent infected over much of their acreage after the second outbreak in 1983.

Subsequent research has shown that the blister rust occurs in "wave years"—infrequent years when moist, cool conditions encourage the fragile spores from the gooseberry to form and infect the pine. In these "wave years" the disease spreads to previously uninfected regions and infects increasing numbers of trees. In nonwave years, infection is low, giving the misleading impression of stabilization. Unless something changes, Sugar Pines will become increasingly rare in most or all of their range.

The genetics of resistance are complex. Some Sugar Pines appear to have total resistance to blister rust, others no resistance. The U.S. Forest Service Institute of Forest Genetics in Placerville has embarked on a plan to locate resistant trees and grow their offspring. This entails finding healthy, potentially resistant pines, collecting their seeds, and sending the seeds to nurseries where they can be germinated. The seedlings are then exposed to blister rust to determine their resistance status and, hopefully, to find resistant ones for planting.

What else can be done? Over the short term, a moratorium on the logging of all healthy, potentially resistant Sugar Pines throughout the state would prevent the loss of a single one of the precious resistant genes to the saw. Abolishing clear-cuts, which create conditions favorable to gooseberry and currant regeneration, would break the rust cycle. Since the rust spores that move from gooseberry and currant to pine are frail and can travel no more than 1,500 feet (450 m), controlling these shrubs would effectively reduce Sugar Pine infections within that range.

Change is natural in a forest, but the loss of a prized native tree to an introduced disease is not a natural change. Like the introduced Chestnut blight and Dutch Elm disease of the eastern United States, blister rust must be fought with all the tools in foresters' arsenals. It will take them all—plus some luck on inborn resistance—to keep the irreplaceable Sugar Pine a viable part of California's midmountain forest scene well into the future.

In addition to the two giant pines—Sugar Pine and Ponderosa Pine— White Fir and Incense Cedar occupy prominent places in Sierra Nevada middle-elevation forests. White Fir, the traditional Christmas tree, surrounds its central trunk with tier upon tier of flat-needled blue-green sprays. Each autumn its resin-covered cones sit upright like candles on the top branches, gradually falling apart on the tree as the seeds are released.

Requiring more moisture than its companion trees and hence more abundant on north-facing slopes, White Fir is usually among the first victims in years of drought. Widespread over many regions of the West, White Fir climbs higher in the western Sierra Nevada than its Mixed Conifer associates, forming an almost pure White Fir zone before it intermingles with the Red Firs (*Abies magnifica*) of the next higher belt.

Incense Cedar has a smaller range of distribution, mostly Californian, with an overlap into mountains of southern Oregon and northern Baja California. But it is probably the most successful conifer in the central Sierra, certainly the most versatile. Able to germinate and grow in both sun and shade, it can outstrip its competitors.

In youth, Incense Cedar trees form broad triangles of dense foliage. Their tiny scalelike leaves press together in short, lacy, flat branchlets which incline in all directions. It is these crushed leaves, along with the terpenes evaporating from the pines, that produce the piney fragrance of the midmountain forest belt on a hot summer day.

The tree itself serves as an effective shelter for wildlife. Its thickly set branches shed rain and snow like an A-frame roof and bear inch-long woody cones (2.5 cm) that drop winged seeds. Mature Incense Cedar's cinnamon-brown bark, fibrous and vertically grooved, offers a warm contrast to the tawny plates of full-grown Ponderosa Pines, the purplish brown trunks of Sugar Pine, and the ashy gray boles of White Fir in western Sierran midmountain forests.

Fire in Mixed Conifer Forests

Neither Incense Cedar nor White Fir were as widespread in the midmountain forests of the Sierra prior to 1900 as they are today. The more pristine Ponderosa Pine forests of early records were open, sunny, parklike places where California Black Oaks shared the spaces with mature Ponderosa Pines, Sugar Pines, ample seedlings of each, and a diverse mixture of shrubs and wildflowers. The midmountain representatives of the manzanita and *Ceanothus* tribes—Greenleaf Manzanita (*Arctostaphylos patula*), white-flowering Deer Brush (*Ceanothus integerrimus*), and blue-flowering Littleleaf Ceanothus (*Ceanothus parvifolius*)—were prominent in the understory, along with Bitter Cherry

(*Prunus emarginata*) and others. Mountain Misery (*Chamaebatia foliolosa*), known as Kit-kit-dizze to the Miwok, covered many lower levels of the forest floor with fernlike, resin-scented foliage; Mahala Mat (*Ceanothus prostratus*) spread matlike over other areas.

Periodic ground fires that swept through the forests, caused by lightning or set by the Indians, burned off the litter and young trees and shrubs but left the fire-resistant large trees relatively intact and the forest mostly open. The shrubs came back, either sprouting from the base or from seeds stimulated by the fire. Ponderosa Pine seedlings got off to a fast start in the sun and nutritious mineral ash. A year-old seedling, 4 inches high (10 cm), looking like a little palm tree with cotyledons coming out from one central top point, already had a taproot 2 feet long (.6 m), and a 4-year-old sapling could tap water from a depth of 5 feet (1.5 m).

Changes occurred when a policy suppressing wildfires came into vogue around the turn of the century. The California Black Oaks and Ponderosa Pines, which need sunlight in youth and favor it in maturity, were gradually overshadowed by invading White Fir and Incense Cedar that sprouted in large numbers in the fire-free forests, often in the shelter of the oaks and pines. As the shade-adapted invaders spread, they blocked out many of the sun-loving shrubs and herbs, along with seedling oaks and pines. In some areas, an understory that could tolerate less light replaced them—Mountain Dogwood (*Cornus nuttallii*), Hazelnut (*Corylus cornuta*), and Thimbleberry (*Rubus parviflorus*), along with thickets of White Fir and Incense Cedar.

It wasn't until the prescribed burns initiated by Richard Hartesveldt, Harold Biswell, and others in the southern Sierra in the 1960s pinpointed the extent of the White Fir-Incense Cedar takeover that the danger posed by these flamable thickets to the life of the entire forest became apparent. Foresters and park managers throughout the state began a program of prescribed burning to try to duplicate, under safe conditions, the natural fires of the past. Yosemite National Park and Sequoia/Kings Canyon National Parks have been burning, to open up forests that once were more open, since 1968, and they will be continuing to catch up with 100 years of combustible buildup for decades to come. Calaveras Big Trees State Park initiated prescribed burns in 1975.

Eliminating doghair thickets of fir is not the only goal. Burns are essential to keep the meadows open. One burn in Yosemite's El Capitan Meadow wiped out thirty thousand invading young trees. The

Miwok regularly set fires in Yosemite meadows in earlier years to encourage the fresh green growth that attracts deer, to maintain open
oak woodlands for abundant acorn harvests, and to prevent ambush by
enemies.

The fires that are natural to Mixed Conifer Forests deal varying cards
to their bird life. Some species enter a burn immediately after a fire, or
even while it is still burning, to exploit suddenly available food. Swifts
and swallows forage on insects at the edge of smoke columns while
Yellow-rumped Warblers pick insects off the hot ash.

Fires attract bark and wood-boring beetles, which in turn draw
Downy, Hairy, White-headed (*Picoides albolarvatus*), and Black-backed
Woodpeckers (*Picoides arcticus*) to devour the beetles. Opportunistic
Steller's Jays join the beetle onslaught and gobble up any insects ex-
posed by the flames. Robins and Chipping Sparrows (*Spizella passerina*)
continue to find worms and seeds in forest openings spared by fire.
Western Wood Pewees (*Contopus sordidulus*) and Townsend's Solitaires
(*Myadestes townsendi*) gain added open space for sallies after flying in-
sects in the burned-out understory.

Birds with different requirements do not fare as well. The loss of
thickets of young trees eliminates preferred nesting sites of Hermit
Thrushes (*Catharus guttatus*), along with the shade and the forest floor
invertebrates which they like. The birds that live by gleaning insects
from leaves decline—Golden-crowned Kinglets, Mountain Chickadees
(*Parus gambeli*), Hermit Warblers (*Dendroica occidentalis*), Western Tan-
agers (*Piranga ludoviciana*). Flickers find their usual steady supply of
forest floor ants smothered by white ash.

The burned-out shrubs that normally provide food, nests, and shelter for Fox Sparrows (*Passerella illiaca*), MacGillivray's (*Oporornis tol-
miei*) and Nashville Warblers (*Vermivora ruficapilla*), and Mountain
Quail (*Oreortyx pictus*) take 5 or more years to regrow to protective size
after fires. But as they come back, they offer an ideal mix of cover and
openness for the handsome, fast-running quail.

The largest native quail in North America, Mountain Quail differ
from others of their kind in their mobility, migrating on foot up the
Coast Ranges and the Sierra Nevada in spring and downward in family
groups in the fall. Male and female are nearly identical in appearance,
gray-breasted with striking maroon and white flanks and throat. Only
the straight plume atop the crown is different, the male's rising higher
than the female's. The male's loud, mellow notes of courtship carry for

nearly a mile through Chaparral and forest in April, May, and June. The quail themselves keep out of sight most of the time, preferring to run from cover to cover rather than fly.

The changes in plant life brought by fire are generally temporary, viewed from the long term, and randomly dispersed. For as fire sweeps through a forest, skipping and jumping local areas, it creates a mosaic of habitats, some more diversified than before. Tall snags killed by fire furnish important home sites for as many as one-third of the birds and mammals in Mixed Conifer Forests. Woodpeckers do most of the initial carpentry, usually excavating new nest holes each spring. Timber drillers such as Williamson's Sapsuckers (*Sphyrapicus thyroideus*), Red-breasted Sapsuckers (*Sphyrapicus ruber*), and Black-backed Woodpeckers prefer hard snags with little decay, whereas White-headed Woodpeckers, Northern Flickers, Hairy Woodpeckers and Red-breasted Nuthatches like soft snags.

Their abandoned cavities become breeding and roosting dens for Mountain Chickadees, Tree Swallows (*Tachycineta bicolor*), House Wrens, Pygmy, Screech and Saw-whet Owls, flying squirrels, Douglas Squirrels, and many other members of the forest community.

Preferences for certain kinds of trees as nesting sites vary widely geographically. In one field study, Northern Flickers chose Quaking Aspen trees (*Populus tremuloides*) in Ontario, Douglas-Fir in British Columbia, Ponderosa Pine in Oregon, Western Larch (*Larix occidentalis*) in Montana, and White Fir in California. The flicker, a fairly weak excavator, selects trees decayed to its liking, wherever it lives, regardless of species.

Forests that have trees in all stages of living and dying, burned and unburned, accommodate the wildlife that utilizes each stage and hence contain the greatest biodiversity. By virtue of their sheer trunk size, however, old growth trees have been found to provide the most nesting cavities, whether in Western Larch–Douglas-Fir Forests near Glacier National Park, Montana, Douglas-Fir Forests of western Oregon, or Mixed Conifer Forests in California.

The most spectacular nester, by far, in California's Mixed Conifer Forests is the crow-sized bird whose loud ringing call and powerful drumming carry its messages long distances through the trees—the Pileated Woodpecker (*Dryocopus pileatus*). When its favored old-growth trees are gone, the big, secretive woodpecker will sometimes opt for clumps of high snags or groups of tall, dense, mature conifers as a nest site. Rich black with white wing patches and a trademark scarlet topnot, the male Pileated excavates a hole 6 to 10 inches deep (15 to

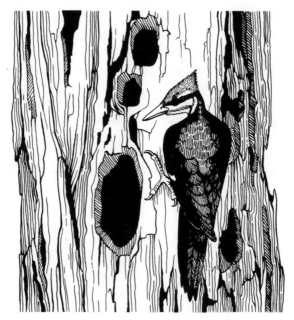

Figure 23. Pileated Woodpeckers excavate large, well-
hidden nest holes high in old-growth or
mature mixed conifers.

25 cm) high in a well-hidden tree. His mate, usually a partner for life, lays 3 to 5 off-white eggs which hatch in about 18 days into naked, long-necked, big-footed babies whose working organs show through transparent skin.

Within 10 days the gawky fledglings are all stomach and mouth, instinctively sucking in the regurgitated ants and beetle larvae which both parents poke down their gullets. As the juveniles grow, they pop up at the entrance hole, rasping loudly, whenever the adults land on the side of the tree with food.

By month's end, the youngsters teeter on the hole edge before falling out and flying off for the first time. They stay together on the home range for several weeks, however, as their parents continue to feed them.

Nonmigratory pairs defend a territory up to a square mile, with both sexes drumming on hollow, resonant trees to assert territorial claims. The blows that their firmly anchored bills deliver pack great power, their heavy head musculature and thick, spongy cranium handling the vibrations with ease.

Bracing themselves on spiny tails, holding on with four toes (two up, one down, and one horizontal), and using the whole body as a hammer from a rear end fulcrum, they make the chips fly as they excavate nest holes big enough for a man's arm to enter or drill to spear juicy Cerambycid beetle grubs with their long sticky tongues.

Nearby nests of Brown Creepers in bark crevices, or nuthatches, flickers, and sapsuckers in cavities, are never molested by the woodpeckers, but their own fledglings occasionally are killed and eaten by reconnoitering Douglas Squirrels. Abandoned Pileated Woodpecker holes, enlarged by weathering and fire, along with recesses in burned-out snags and logs, provide homes for Raccoons, Great Horned Owls, Martens, and Fishers, new shelters made available largely by fire in the midmountain forests.

Root Rot and Boring Beetles

Where Sierran mixed conifers have escaped fire over long intervals and have grown too close together for their own health, another forest agent initiates major ecological changes of its own. Annosus Root Rot (*Heterobasidion annosum*), formerly known as *Fomes annosus,* is a fungus which attacks conifers that grow in dense clusters. Annosus is spread by windborne spores to fresh stump surfaces or wounds on the base of trees. When the spores germinate, the fungus grows along the roots, killing them. Where diseased roots contact those of a healthy tree, the fungus spreads to that tree.

Hence, from an initial invasion of a forest, trees die in a circle around the original infection, and new circles continually form and spread. Since the disease especially affects the large structural roots, it weakens a tree's ability to stand, and green trees topple unexpectedly in windstorms, or even on calm days. Warning signs are few—sometimes just brown branches or a flattened top.

Thousands of Yosemite Valley conifers have fallen from the disease, which, although native to the state and worldwide in distribution, was not identified in the park until the early 1960s. Decades of field research by John R. Parmeter, Jr., and others, have disclosed how Annosus spreads. His studies cite heavy development in Yosemite Valley and the bark beetle as major contributing factors. These factors resulted in many trees being cut down and left as stumps, and unbeknownst to park managers, the stumps served as foci of infection for the root rot.

From diseased stumps or trees, the fungus progresses through stands of timber at roughly 3 feet a year. At this rate, if infection centers are numerous, Annosus can virtually wipe out a coniferous forest within a few decades. This is happening in Yosemite Valley! Nearly one million board feet of infected trees have been removed from Yosemite Valley alone. There are no known ways to stop the root rot's spread. Annosus root rot, along with fire, drought, wind, bark beetles, and fungi, is just one more natural agent by which Mixed Conifer Forests prune and reshape themselves over the centuries.

Beetles exert their own major environmental influence on mountain forests. Over 300,000 species strong on the planet, they comprise the largest order of insects. And at least 7,000 kinds of them live in California. Some of their richest habitats are under the bark of snags, stumps, and living or fallen trees.

Campers in Mixed Conifer Forests are sometimes startled on warm, summer nights when a huge brown beetle, more than 2 inches (50 mm) in length, bangs into their lanterns or RV windows. The Spiny Wood-borer or Pine Sawyer (*Ergates spiculatus*), one of the two largest beetles in the West, is attracted to lights and hits them with the impact of a small bird.

The damage it does to trees comes from eggs the adult lays in bark crevices of conifers, especially Ponderosa and Sugar Pines and Douglas-Fir. The eggs hatch into cream-colored, grublike larvae with reddish brown heads and powerful mouthparts. The larvae chew galleries through the sapwood deep into the heart of the tree, growing thick bodies almost 3 inches long (70 mm). Loggers know these "timber worms" well.

The larvae live a number of years and occasionally turn up in furniture or lumber long after their tree hosts have been cut. When they mine conifers killed in a forest fire, they speed up the deterioration of the wood and reduce the salvage possibilities. Ponderosa Pines weakened at the base by Spiny Wood-borers may fall unpredictably, especially when bark beetles add their input.

Another large beetle that lives under the bark of conifer stumps gives hikers who come across it several distinct surprises. The beetle appears to have eyes on the front part of its upper side (prothorax), large velvety black ovals encircled in white. They are only colorations. When lying on its back, the beetle can suddenly spring an inch or two into the air with a loud click.

This largest of the click beetle family in California, the Eyed Elater (*Alaus melanops*), measures up to 1 1/2 inches (40 mm). It produces even longer yellowish larvae which live in logs and stumps and prey on grubs of other wood-boring beetles.

Mixed Conifer Forests North and East

Soil moisture is a critical factor in determining which conifers grow in the Mixed Conifer Forests in different parts of California. The midelevations of the Sierra north of Lake Tahoe receive considerably more rainfall than regions south, and their cool, moist sites host some of the plants of similar climates in the northern Coast Ranges. Douglas-Fir grows prominently there, along with scattered Pacific Madrone, Tan Oak, occasional California Nutmeg, and Pacific Yew.

Douglas-Fir, which attains good size on moist, north-facing sites as far south as Yosemite, ends its Sierran range 20 miles south of the Park. Pacific Yew reaches its southern Sierran limit in the Giant Sequoia groves of Calaveras Big Trees State Park, the only place where Pacific Yew and Giant Sequoia occur together. Nutmeg thrives at the Arch Rock entrance to Yosemite, as well as along some of the boulder-lined Park roads.

The eastern side of the Sierra, steeper and drier than the west, carries poorly developed Mixed Conifer Forests. South of Sonora Pass, White Fir and Jeffrey Pine (*Pinus jeffreyi*) occupy most stands. Jeffrey, which replaces Ponderosa Pine in higher, colder sites on the west Sierran slope, forms nearly homogeneous stands on the east side near Mammoth Lakes.

To the north, Jeffrey Pine grows intermixed with Ponderosa Pines on the east side of the Cascades in Lassen National Forest. And in the far northeastern corner of California, where the Warner Mountains shoot up like a little Sierra Nevada amid Great Basin Sagebrush, Jeffrey Pines join the Warner Mountain forests of Quaking Aspen, Ponderosa Pine, White Fir, Western Juniper (*Juniperus occidentalis* var. *occidentalis*), and the rare, little-known Washoe Pine (*Pinus washoensis*).

The Washoe Pine's possible kinship to Jeffrey or Ponderosa Pine has long intrigued botanists. Discovered in 1938 on the east slopes of

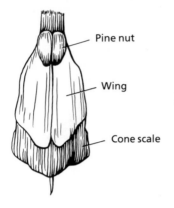

Pine nut

Wing

Cone scale

Figure 24. The woody scales that form a pine cone each contain two pine nuts at their inner base. When the pine cone opens, the nuts drop out and float to earth on their wing—a thin parachute that twirls them about and may carry them away from the parent tree.

Mt. Rose, Nevada (northeast of Lake Tahoe), where it grows in mixed stands with Jeffrey Pines, Washoe resembles Jeffrey Pine from a distance. It has the familiar reddish brown bark, needles in threes, and nearly identical growth habit. Washoe Pine cones, almost exact mini-versions of Jeffrey's, show similar slender, down-pointing prickles on the cone scales but are just half the Jeffrey cone's 5 to 10-inch (12 to 25 cm) size. There the resemblance stops.

Every species of pine possesses its own distinctive wood resin chemistry. Washoe Pine differs strongly from Jeffrey Pine in the composition of the turpentine fraction of its resin. It does hybridize with Jeffrey, but only at a very low rate of success. Chemically, Washoe is closer to Ponderosa Pine, hybridizing rather readily with Ponderosa's Rocky Mountain race.

Ponderosa Pine, which grows on sunny mountain slopes in every western state, has diversified over time into eastern and western varieties, with several geographic races in each, thus encompassing a wide genetic range. *The Jepson Manual* calls the California trees Pacific Ponderosa Pines. Washoe Pine may be a Pleistocene offshoot of one of the ponderosa races, isolated from its parents long enough to develop genes uniquely its own. Its exact status, however, remains a botanical guess.

In the few scattered locales of northeastern California and Nevada where Washoe Pines occur, both Jeffrey and Ponderosa Pines are regularly present or nearby. Ponderosa and Jeffrey are themselves such closely related yellow pines that they were once considered varieties of the same species. The differing chemistry of their resins plus the size and structure of their cones separate them distinctly. In the field, Jeffrey

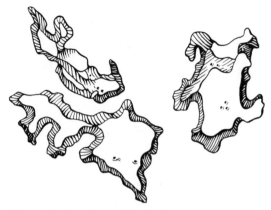

Figure 25. Ponderosa Pine bark on mature trees
frequently breaks up into tawny pieces
shaped like parts of a jigsaw puzzle.

Pine cones are much larger and less prickly to handle. Where their
ranges overlap, the two species sometimes hybridize, but Jeffrey Pines
generally tolerate drier soils than ponderosas. Side by side, ponderosa's
needles stand out greener against Jeffrey's grayer ones. Young branches
of Jeffrey Pine are lighter in color and smoother than Ponderosa Pine
branches. Ponderosa buds are sticky, and only Jeffrey Pine gives out the
fragrant aroma of pineapple, vanilla, or banana from the cracks in its
bark on warm summer days.

In the Warner Mountains, where Jeffrey and Washoe Pines grow on
slopes near Blue Lake, at an elevation of 6,000 feet (1800 m), the wildlife in action on a summer day could easily be that of most Mixed Conifer Forests in California.

A chipmunk sitting atop a volcanic boulder, tail twitching from
side to side, drops to the forest floor to dig beneath silvery-leaved lupines. Golden-mantled Ground Squirrels (*Spermophilus lateralis*) scamper across a clearing and disappear into burrows with mouthfuls of pine
needle mulch. The hoarse song of a Western Tanager chops out rapid
phrases from a Ponderosa Pine branch, where it fluffs its red head and
luminous yellow rump before flying to the lake shore for a drink. Deer
and Raccoon tracks dent the wet sand. From halfway up a White Fir, a
Douglas Squirrel scolds, its tail jerking spasmodically over its back in
time with the notes, while high above, on the fir's broken top, a Northern Goshawk intently scans the scene.

This Mixed Conifer Forest of the Warner Mountains shows a distinct affinity to the Great Basin surrounding it. The presence of Western Junipers, Big Sagebrush (*Artemesia tridentata*), and Rubber Rabbit-brush (*Chrysothamnus nauseosus*) cements that Great Basin tie over the 80-mile length of the range.

Mixed Conifer Forests of Southern California

The midmountain forests of southern California share the same major conifers as western Sierran forests. Ponderosa Pines cover the lower, more mesic slopes from 4,500 to 7,500 feet (1,350 to 2,250 m). Jeffrey Pines replace them on the colder, higher sites at 5,000 to 9,500 feet (1,500 to 2,850 m), with a few hybrids where they overlap. Moist areas hold mature Sugar Pines, White Fir, Incense Cedar, and Mountain Dogwood amid a shrub layer resembling the Sierran. Canyon Live Oaks and Interior Live Oaks mingle with the conifers.

Much of the bird life is similar. Acorn Woodpeckers drill in the California Black Oaks. The descending melody of the Ruby-crowned Kinglet, the "dee-dee-dees" of Mountain Chickadees, the harsh scolding of Steller's Jays, and other familiar montane bird songs resound through the mixed forests.

The southern California forests differ from Sierran forests of the same type in two prominent ways. First, the southern forests always seem to have a big hill of Chaparral and rock just around the bend; secondly, the southern California forests provide the major home of Coulter Pine and the only home of Bigcone Douglas-Fir. These two conifers occupy the transition zone between lowland Chaparral and the Mixed Conifer Forests above.

Coulter Pine (*Pinus coulteri*) is a medium-sized, open tree with tufts of long blue-green needles in bundles of three. Up to a foot in length (30 cm), the needles, stiff and three-sided, form dense clusters. Growing chiefly in or above Chamise and manzanita Chaparral on ridges and south-exposed rocky slopes, Coulter Pine tolerates heat and drought.

The tree's cones are truly impressive, the most massive of any pine cones in the world. Armed with strong, incurved hooks on thick cone scales, and dripping with resin, they weigh up to 9 pounds (4 kg) each. Some cones reach 14 inches (35 cm) in length.

Sharing many of the same intermediate mountain elevations as Coulter Pine but utilizing a totally different habitat, Bigcone Douglas-Fir (*Pseudotsuga macrocarpa*) rises from the moist, north-facing slopes and canyons of southern California's mountains. Growing on the cooler, steep inclines from the Santa Inez Mountains to the Cuyamacas in San Diego County, it ranges through 2,200 to 7,000-foot elevations (600 to 2,100 m).

Even a ridge away, Bigcone Douglas-Fir's silhouette stands out. Long horizontal branches stretch sideways to interlace with those of its neighbors, forming continuous dark, outreaching lines. Younger trees, with soft foliage and drooping twigs, resemble their northern cousin, Douglas-Fir. But the mature trees attain none of the Douglas-Fir's enormous size, averaging a mere 30 to 60 feet tall (9 to 18 m).

The cones of each confirm their close relationship. Bigcone Douglas-Fir cones possess the typical three-pronged bract of the Douglas-Fir, but the bract does not extend as far beyond the cone scales and the cones are much bigger, 4 to 8 inches (10 to 20 cm) long. Neither the stem-encircling needles nor the papery brown cones add enough weight to bend the straight-out limbs at their tips.

Bigcone Douglas-Fir and Coulter Pine both grow intimately with Canyon Live Oak, but, because of their different habitats, they have dissimilar ecological relationships with it. Canyon Live Oak takes on many growth forms, from shrub to tall tree. Repeated burnings reduce it to a shrub. Where the oak grows in the Chaparral surrounding Coulter Pines, the fires that frequent the region normally burn both oak and Chaparral completely to the ground and kill the pines.

The oak resprouts from the crown just above its root-mass into a round shrub at the same time that the Chaparral is resprouting and re-seeding. For its comeback, Coulter Pine, though not a closed-cone pine, utilizes a similar fire-adapted strategy. Its heavy cones open in mid-winter, well after the fire season. If a fire has gone through, Coulter's seeds drop onto a nourishing mineral ash bed. The seedlings, remarkably drought resistant, send down deep roots the first year and mature quickly, bearing cones within 10 to 15 years, often before the next fire. Repeated fires, however, favor an eventual manzanita takeover, forcing Coulter Pine uphill into areas of less Chaparral competition.

In contrast, Bigcone Douglas-Fir's forests are long-lived, occupying precipitous slopes and sheltered canyons less prone to fire. Here their seeds reproduce most successfully in the shade of tall Canyon Live Oak trees and little brush. Although the ruggedness of the terrain precludes most fires, Bigcone Douglas-Fir is the only southern California conifer capable of sprouting from large branches and from the trunk after defoliation by flames.

Mixed Conifers—Santa Lucia Mountains

A distinguished variation of the state's Mixed Conifer Forests graces the Santa Lucia Mountains south of Monterey. Rising out of Redwood-lined creek and river valleys just inland from the Pacific Ocean, the Santa Lucias carry a unique mixture of coastal hardwood trees and Sierran conifers, along with a rare bonus entirely their own.

Coast Live Oaks and Pacific Madrones dominate many of the lower north slopes and canyon bottoms, replaced higher up by Tan Oaks and Interior and Canyon Live Oaks. Other cool, moist canyons house Incense Cedar. Good stands of Ponderosa Pine, California Black Oak and Sugar Pine grow scattered over the range, with understories of Sierran shrubs and forbs not found elsewhere in the southern Coast Ranges. Amid the Chaparral grow a few Jeffrey Pines, along with Coulter Pines, some Coulter-Jeffrey hybrids, and Knobcone Pine.

But the prize of the Santa Lucias is the rarest and most unusual fir in the world—the Bristlecone or Santa Lucia Fir. Climbing the steep, rocky slopes and ridges at 2,000 to 5,000-foot elevations (600 to 1500 m), Bristlecone Firs (*Abies bracteata*) give the range its distinctive botanical appeal. Imposing conifers, the firs reach heights of 150 feet or more (50 m). From a wide-bottomed pyramid with branches close to the ground, they taper to a narrow crown and spirelike top.

The stiff, 2-inch-long (5 cm) needles, exceptionally long for a fir, end in extremely sharp points. Dark green above and whitish below, they attach to the twigs with twisted petioles, resulting in flat sprays of foliage.

The purple-brown cones, which give the tree its name, grow in the tips of the tree's crown and are special in possessing needlelike bracts that curve out among the cone scales. The cones produce abundant seeds, but these frequently turn out nonviable because of parasitism by a minute orange and brown wasp.

Called a seed chalcid (*Megastigmus* spp.), this tiny female wasp inserts the spearlike ovipositor at the tip of her abdomen into the fir's seeds at a time when the young cones are soft and vulnerable. The larva that hatches there eats the inner part of the seed, effectively destroying it. A neat hole in the outer seed coat remains the only clue to the insect's later emergence. Enough seeds escape, however, to provide continuing healthy Bristlecone Fir replacements in this geographically restricted Mixed Conifer Forest of the southern Coast Ranges.

Mixed Conifers and Air Pollution

It was in the Mixed Conifer Forests of the San Bernardino Mountains east of Los Angeles that air pollution damage to Pacific Ponderosa Pines first became recognized in the early 1960s. For nearly a decade a mysterious blight known as "Disease X" had been turning the needles of mature Ponderosas a mottled yellow before destroying the trees. No one guessed, at that time, that the smog from the Los Angeles Basin 60 miles away (96 km) could be the killer.

As trees died by the thousands, Paul Miller and colleagues traced the lethal agent to its source. The brown smog that blankets Los Angeles comes largely from millions of automobiles. When hydrocarbons and nitrogen oxides from car exhausts combine in sunshine, one of their major products is ozone (O_3). An invisible, eye-stinging, caustic gas, ozone can crack rubber, deteriorate fabrics, scar lungs, and cause coughing, shortness of breath, pain, and fatigue. And, it kills trees!

Ozone, along with all the other air pollutants generated in Los Angeles, ordinarily remains trapped there during much of the day by inversion layers of warm air over the mountains rimming the basin. When afternoon offshore breezes blow the smog-laden air upslope into the mountains, ridges in direct line with the air currents suffer devastating tree losses.

Ozone acts quickly. Ponderosa Pines usually keep their needles 3 to 4 years. Ozone-struck trees shed all but the current crop, leaving a sparse, stripped tree. The remaining needles soon show yellow mottling as ozone destroys the chlorophyll. As the needles fall, the root system deteriorates and resin flow in the trunks slows down, opening the way for bark beetles.

Fortunately, not all kinds of trees succumb in the same degree and, even among the most sensitive Ponderosa and Jeffrey Pines, some specimens show an inborn resistance. Sugar Pine, luckily, seems relatively immune. But the list of vulnerable conifers grows with each decade of bad air exposure. And this disaster is not confined to southern California.

Ozone damage, now known as ozone mottle, became visible in the southern Sierra in the mid-1970s and continues to increase ominously. On countless days a brown layer of pollution hangs over the San Joaquin Valley west of Sequoia/Kings Canyon National Parks. It shows up plainly from Moro Rock. Afternoon upslope breezes blow the ravaging pollutants into the 5,000 to 7,000-foot levels (1,500 to 2,100 m) of the parks, where ozone works ruinous havoc on the pines and oaks and on some Giant Sequoia seedlings. Sequoia National Park has recorded the highest cumulative levels of ozone over a one-day period for all of the national parks. Levels in Sequoia regularly climb higher than those of Los Angeles.

Yosemite National Park, farther north, suffered a fivefold increase in ozone damage between 1985 and 1990. Thirty percent of its Jeffrey and Ponderosa Pines show yellow needle mottling, and no area in Yosemite's mixed conifer belt stands free from ozone's relentless scourge. Adjacent national forests display steadily growing numbers of Ponderosa and Jeffrey Pines with the thinning crowns and mottled needles that are ozone's trademark.

California's Great Central Valley sits in the midst of an even larger basin than Los Angeles, bordered by mountains and hemmed in by the same inversion layer that traps smog beneath. As populations of valley cities boom, more and more of their polluted air follows its daily, deadly flow uphill to the Mixed Conifer Forests of the western Sierra Nevada.

The most diverse coniferous forests on this earth, still very beautiful, face all the natural ecological challenges of forest life—fire, drought, insects, fungi, winds—with adaptations built in over centuries. Their genetic resistance to poisonous air is now being sorely tested. In both the short and the long run, air pollution of human derivation will require a solution for humans and trees, for both are dependent on the same air for survival.

8

Giant Sequoia Groves

Amid the Mixed Conifer Forests of the western Sierran slope, in certain special places, stand groves of the largest living tree on our planet, the Giant Sequoia or Big Tree (*Sequoiadendron giganteum*). Visualize trees 30 feet across (9 m) that would fill an average city street and block out buildings across the way; trees 300 feet high (90 m) that, lying on their sides, would cover a stadium football field from goal line to goal line; trees that take forty-five paces to walk around. These are the monarchs of the tree world.

Never occurring in pure stands, Giant Sequoias, nonetheless, dominate the forests where they live, both in beauty and in size. Their companion Sugar Pines, White Firs, and Incense Cedars pale to insignificance alongside the immensity and majesty of the Sequoias.

Seventy-five groves of Giant Sequoias lie scattered over a narrow 260-mile-long belt (416 km) in the southern and central Sierra Nevada at an elevation of 4,500 to 7,500 feet (1,350 to 2,250 m). Most of them, including the largest groves, Redwood Mountain and Giant Forest, occur in the southern one-third of their range, in or near Sequoia/Kings Canyon National Parks.

The largest individual trees include the General Sherman, Lincoln, and Washington trees of Sequoia National Park, General Grant of Kings Canyon National Park, the Boole tree of Sequoia National Forest, and the Grizzly Giant of Yosemite National Park's Mariposa Grove. General Sherman gets top billing as the most massive living tree on earth, 36 feet across (11 m) at the base and 275 feet high (83 m),

Map 8. Giant Sequoia groves.

totaling a volume greater than any other known. Like many veteran sequoias, Sherman rises branchless for a good part of its height, growing straight, with barely a taper. In 1978 the General dropped a branch which was, in itself, larger than any tree east of the Mississippi River, 150 feet long (45 m) and nearly 7 feet in diameter (2 m).

Other Giant Sequoias grow taller than General Sherman, topping 300 feet (90 m) but are not as broad. No known Giant Sequoia, however, approaches the record of the tallest Redwood on California's northern coast, 368 feet high (110 m).

It is not just the huge size of the Giant Sequoias that makes their groves such impressive places. The groves possess an ambience all their own. Moist, cool retreats, they are often shallow basins where heavy snows bed the trees down in winter and feed an underground water supply which their roots tap during the dry summers. In bright sunlight, the sequoias' cinnamon trunks soar straight up into clouds of thick green foliage. In fogs the giant boles loom dark and formidable,

Figure 26. White-headed Woodpeckers, permanent residents of
Giant Sequoia groves, often nest in the Sequoia's soft,
thick cinnamon bark.

with all blotted out behind. In back light, close-growing groups turn
chocolate brown, their bases separated from each other only by sunlit
edges. Russet-red tones pervade the groves—rufous barks, reddish
brown earth littered with bronzy cones, coppery needles, dried scaly
leaves, pink-barked branches broken off in winter storms.

In addition to the colors of these select places, there are the sounds.
The same birds and mammals of the Mixed Conifer Forests live among
the Giant Sequoias. The scold of the Douglas Squirrel, the trill of
Dark-eyed Juncos, the ringing call of Pileated Woodpeckers, the mellow
song of Black-headed Grosbeaks all liven summer mornings. And in
winter, the lispy notes of Golden-crowned Kinglets and Brown Creepers and the sharp rattle of White-headed Woodpeckers echo through
the groves.

The silent times are broken only by the rustle of seeds and tiny twigs drifting slowly to the forest floor—seeds whose ancestors date back 175 million years, seeds falling from trees that individually span forty human lifetimes.

And even when a tree is 3,000 years old, it continues to grow. In fact, Giant Sequoias rank among the fastest growing trees in the world. During their first 5 years they rise quickly to an average man's height, clothed to the ground with scalelike leaves. By age 10, young Sequoias attain 20 feet (6 m), their pointed spires shedding snow like an A-frame cabin.

Adding up to 2 feet a year, they top 100 feet (30 m) by their fiftieth birthday. From then on, as the trees grow taller, they begin to self-prune their lower branches, dropping them after drought or after fires destroy surface roots. Beyond 100 years, Giant Sequoias gradually take on the rounded dome of the mature tree and continue adding growth rings as they climb to 200 to 300-foot heights (60 to 90 m).

Many old giants develop snag-tops, where parts of the crown die back because of lightning strikes or fire damage. But even these trees go on adding girth, about one foot per century. Because of their immense size, this increases amounts to enough wood for a new tree 60 feet tall (18 m) and 1 1/2 feet wide (.5 m) every year.

And the reproductive urge never abates. Each winter and spring the ground beneath the old-timers turns golden with pollen shed by the tiny male conelets high in the outer branchlets. Enough of it lands upon the minute female conelets to load down the crowns with around 2,000 new cones every year.

The egg-shaped cones produce chlorophyll that turns them bright green from the beginning, and they quickly develop fleshy cone scales around a central core. Mature in 2 years, the cones remain green and attached to the tree for roughly 20 more years, enlarging gradually all the time. The cone's stalk adds a growth ring each year, countable under a microscope, a miniversion of the annual ring on the trunk. Cone size averages 2 1/2 inches long (6 cm), showing a genetic variation from tree to tree. Cones with viable seeds appear on some Giant Sequoias as early as 6 years after germination, on trees only chest high. But most Sequoias produce cones much later. Around 150 to 200 seeds remain trapped inside the tightly closed cones during their two decades on the tree, unless certain actions release them.

Those actions were first revealed in the late 1960s by a team of biologists from San Jose State University—Richard Hartesveldt, Tom Harvey and colleagues—working in Redwood Mountain Grove. Rigging an elevator to climb high into the canopy, they were able to investigate life in the 200 to 300 foot level (60 to 90 m) of Giant Sequoias for the first time and to follow it on the ground below.

The most readily observed releaser of Sequoia seeds turned out to be the sprightly Douglas Squirrel, commonly known as Chickaree in the Sierra Nevada. The Pacific Coast version of the eastern and northern Red Squirrel, the Douglas Squirrel is just as hyperkinetic in action. In pine and fir forests it chews the dry woody scales of cones down to the nubbin to get the pair of nuts lodged there, leaving the ground littered with skeletal brown cores, nut husks, and loose scales.

But the tiny dry seeds of Giant Sequoias do not tempt the Douglas Squirrel. Instead, it relishes the fleshy, thick green scales of the younger Sequoia cones. In stripping this green flesh, it dislodges many seeds. Eating Giant Sequoia cones on the tree, in territories it maintains the year around, the squirrel sends seeds earthward at all seasons and also cuts and stores large numbers of cones for winter use. In autumn it often eats young green cones like corn-on-the-cob, leaving the ends intact.

Its incisors can nip off tender green cones at an amazing pace, over 500 in 30 minutes at peak performance. Burying the cones by the hundreds in cool, wet places, the little dynamo spills many seeds near the cache when it returns later to eat the flesh. Numbers of these seeds germinate, if the site is favorable.

Since Douglas Squirrels find the green cone scales edible only while the flesh is still young and tender, they limit their cone choices to those that are 2 to 5 years old. Between the ages of 4 to 9, the cones in the top half of the tree face another attacker, a tiny long-horned beetle (*Phymatodes nitidus*). Until 1968, this little wood-borer's role in Giant Sequoia regeneration was totally unknown.

The female cone beetle lays her eggs in the junctions of the cone scales. They soon hatch into larvae about 1/6 inch long (4 mm), which chew their way into the cone's interior. As they tunnel, feeding on the cone's tissues, they often sever the xylem pipeline to the cone's scales. This cuts off the water supply and causes the fleshy green scales to turn brown and shrink, creating cracks through which the tiny seeds tumble out.

Of the nearly 40,000 cones that hang in the crown of a mature Giant Sequoia at any one time, roughly one-third qualify as brown, beetle-invaded cones. Tiny exit holes furnish proof that the beetle was there,

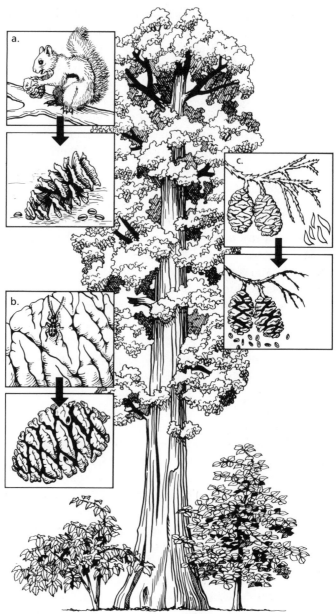

Figure 27. Giant Sequoia seed dissemination. (a) Chickaree eats fleshy green cone scales, releasing seeds. (b) Cone Beetle lays eggs in cone. Beetle larvae cut water line, causing cones to open and release seeds. Adult beetle escapes through hole. (c) Fire updrafts dry and open cones, releasing seeds.

benefitting by a reciprocal arrangement: the Sequoia's cones furnish food for the beetle; the beetle's actions help release seeds from the tree.

Cones also fall during winter storms, during strong winds, or from heavy loads of snow and ice, and shed their seeds wherever they happen to land.

And then there is fire. Hot updrafts from fires rise through the foliage to dry and open the cones. A week or two later, thousands of seeds rain onto the newly burned, friable mineral ash below.

Getting seeds to the ground by any of these means is only part of the journey to a new tree. Living as long as they do, Giant Sequoias need only succeed in replacing themselves once in 1,000 to 2,000 years to keep their numbers constant. But accomplishing that requires a certain amount of luck.

Giant Sequoia seeds are so small, one-eighth of an inch (3 mm), and so lightweight, 91,000 to a pound, that to survive they must make contact with earth. Freshly burned soil, leached by rains, usually has temporary spaces between soil particles which serve the tiny seeds well. Disturbed ground along troughs where trees have fallen also generates rows of seedlings. Seeds that land on ice and snow may germinate after the melt.

If the requisite soil moisture and sun are present and the seed germinates, the seedling moves into the most critical two years of its life. Because so little food is stored in Giant Sequoia seeds, each new seedling must become self-sufficient in a hurry. The tap root must tunnel down to water and spin off laterals which will take over the tree's water supply. Laterals may form without a tap where ample moisture is available. Stem and leaves must shoot up and photosynthesize. And all this must happen quickly, for there is no time to waste in the warm, sunny, summer growing season of the Sierra Nevada.

If the water table dries out to a foot below the surface, and the new seedlings' roots have only penetrated 8 inches (20 cm), fate writes *finis* for the young plants. In addition to desiccation, they can be nibbled off by mammals, birds, or insects, buried by fallen branches, blanched by heat canker, broiled in char-darkened soil absorbing temperatures to 157°F (69°C), have their leaves cemented together by gray mould blight from prolonged burial under snow, grow twisted and contorted in deep shade, or face still other hazards.

Members of the San Jose State University study team tracked over 2,000 Giant Sequoia seedlings that had germinated naturally after a burn. More than half died the first summer. By the end of the second summer, 98.6 percent were dead.

Fire

The lightning-triggered fires that inflamed Sierran mid-mountain forests over the centuries, roughly one every 10 years, as shown by fire scars, left charred reminders in Sequoia groves along the way. John Muir and his mule, Brownie, witnessed a southern Giant Sequoia forest fire in the autumn of 1875 and stayed with it for days. Muir never forgot its impact and described it in vivid detail.

"In the forest between the Middle and East forks of the Kaweah, I met a great fire, and as fire is the master scourge and controller of the distribution of trees, I stopped to watch it and learn what I could of its works and its ways with the giants.

"It came racing up the steep chaparral-covered slopes of the East Fork Cañon with passionate enthusiasm in a broad cataract of flames. . . . But as soon as the deep forest was reached the ungovernable flood became calm like a torrent entering a lake, creeping and spreading beneath the trees where the ground was level or sloped gently, slowly nibbling the cake of compressed needles and scales with flames an inch high, rising here and there to a foot or two on dry twigs and clumps of small bushes and brome grass. Only at considerable intervals were fierce bonfires lighted, where heavy branches broken off by snow had accumulated, or around some venerable giant whose head had been stricken off by lightning" (Muir 1901:307).

Muir observed the fire attack the large Giant Sequoias only at ground level, consuming the fallen leaves and humus at their feet, doing them but little harm unless considerable quantities of fallen limbs happened to be piled about them. The trees' thick mail of spongy, unpitchy, almost unburnable bark afforded strong protection.

Well-planned prescribed burns of today attempt to recreate the mosaic pattern of such wildfires. The flames follow designated paths, eliminating the most combustible debris and reducing stands of White Fir that restrict young Sequoia growth. Islands of vegetation of different ages remain untouched, oases whose flora and fauna eventually repopulate the burn.

In Giant Sequoia groves, good burn preparation includes removal of fallen limbs and debris from the bases of the big trees. Hot fire there might ignite and kill the roots paralleling the surface a few inches down, and might scorch the striking cinnamon barks, diminishing the natural beauty of the scene, as it has in some parks.

Plate 1
The soaring trunks of the Redwood
dominate the luxuriant northern
California forests.

Plate 2
Among the flowers of the Redwood
Forest floor grow (a) tiny exquisite
Calypsos, usually no more than 6 inches
high, (b) Fairy Bells, (c) Redwood Sorrel
with clover-like leaves, and (d) False Lily-
of-the-Valley, spreading in mats over
decaying logs.

Plate 3
The heavy rains in California's North Coastal Forests promote lush growth of (a) Western Hemlock, (b) Western Red Cedar, and (c) Sitka Spruce. (d) Roosevelt Elk roam the forests and meadows.

Plate 4

Redwood, Coastal, and Douglas-Fir/Mixed-Evergreen communities of northern California share many of the same plants and animals. (a) Madrone, (b) Tan Oak, and (c) Pacific Yew fill in middle heights under the conifer canopy. (d) The Banana Slug glides over the forest floor, breathing through the hole in its side, seeing through eyes which are black dots on the tips of the larger tentacles.

Plate 5
Closed-cone Forests feature cones like these of (a) Monterey Pine, which require intense heat to open. (b) Bishop Pines, enshrouded with Lace Lichen, grow in fog-bound coastal areas.

(c) Torrey Pines spread on ocean bluffs north of San Diego, and (d) California Rose-Bay brightens the Pygmy Forests of Mendocino with its May blossoms.

Plate 6
(a) Foothill Woodland often occupies steep hills above river canyons. (b) Blue Oaks form woodland, with or without pines, and house (c) the Scrub Jay as a permanent resident. (d) Globe Lilies liven foothill slopes in spring.

Plate 7
Sierran midmountain forests are colorful in all seasons. (a) Ponderosa Pines, with orange and black bark, stand out in winter snow. (b) In summer, Black-headed Grosbeaks join the (c) resident Western Gray Squirrels and (d) nocturnal Raccoons.

Plate 8
(a) Rare Washoe Pines of the Warner Mountains are a midmountain forest species.
(b) Washoe Pine cones, looking like tiny Jeffrey Pine cones, have a resin chemistry closer to that of Ponderosa Pines.
Southern California's forests feature (c) Bigcone Douglas-Fir with outward reaching branches and (d) Coulter Pine, whose massive cones come armed with stout hooks.

Plate 9
(a) The President Tree in Sequoia National Park is one of many Giant Sequoias towering above minute humans. (b) Lightning fires over the centuries have left charred reminders in Giant Sequoia groves. If enough cambium survives, the tree continues to thrive, despite a burned-out interior. (c) Prescribed burns remove combustible debris and add rich ash to the forest floor, preventing the buildup of dry fuel that could trigger a holocaust.

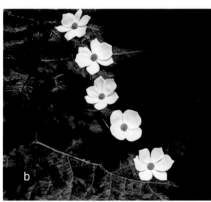

Plate 10
(a) A Giant Sequoia grove emanates russet-red tones the year round. In late spring the colors are enhanced by the floral display of (b) Mountain Dogwood and wildflowers such as (c) Striped Coral-root, hosting a mosquito on its tip, and (d) the Phantom Orchid.

Plate 11
(a) Red Fir Forests form the snow reservoir of California's mountains. (b) In summer, Snow Plants add crimson highlights to the forest floor. (c) Lodgepole Pines often line the shores of high lakes. (d) Aspen groves bring the colors of autumn to high mountain forests.

Plate 12
(a) The Douglas Squirrel or Chickaree fills mountain forests with aggressive calls in defending its territory. (b) Steller's Jay oversees most happenings. Two common rodents thrive in the Red Fir-Lodgepole Pine Forests. (c) The larger, chunkier Golden-mantled Ground Squirrel, with an unstriped copper-colored head, feeds at ground level. (d) The chipmunk, smaller and more delicately built, with stripes on its pointed head, often climbs trees.

Plate 13
(a) Subalpine Forest of Lodgepole and
Whitebark Pine climbs the slopes at
10,000 feet on Tioga Pass in Yosemite.
(b) Whitebark Pine, the dominant timberline tree in most of California's
mountains, survives both as a multi-
trunked tree and as a horizontal mat. (c)
Clark's Nutcracker has a special relationship with Whitebark Pine, feeding on its
nuts and helping to spread them. (d)
Mountain Hemlocks favor the snowy,
cold exposures of north-facing canyons.

Plate 14
(a) Yellow-bellied Marmots den in the rock slides of high mountain areas.
(b) Southern California's subalpine realms include Limber Pine as a dominant tree.
(c) The subalpine heights of eastern California's White Mountains feature the oldest trees on the planet—Western Bristlecone Pines. Many bristlecones endure for over 4,000 years. (d) Even after death, they may stand for 2,000 years more!

Plate 15

(a) Pinyon pines and junipers occupy eastern California's lower foothill slopes, sometimes growing together, sometimes separately. (b) At dawn, in early spring, Sage Grouse cocks strut on their traditional sagebrush mating grounds. (c) In late summer and autumn, Rabbitbrush brings a brilliant gold to eastern California. (d) Mule Deer come down from the higher mountains at the same time.

Plate 16
(a) Thirteen kinds of conifers grow in
the Cedar Basin Research Natural Area
of the Klamath Mountains, a sample of
the region's diversity. (b) From ridgetops
above rugged Klamath canyons, hikers
can view Mount Shasta. (c) At Lake
Eleanor, Deer Oak, one of the world's
rarest oaks, forms thickets above the
azaleas and willows at lake shore. (d)
Seeps throughout the Klamath region
display the green hooded leaves of the
insectivorous California Pitcher Plant or
Cobra Lily. Insects crawl inside the
hood, become trapped, and fall to their
doom.

In October 1988, when the huge Buckeye Fire came roaring up the steep Chaparral-covered slopes in Sequoia National Park, headed for the Giant Forest grove of Giant Sequoias, it was stopped in a buffer zone created by prescribed burns. Park service ecologists, who had anticipated just such a fire, had done their work so thoroughly that many of the Giant Forest border areas would not even carry a backfire.

The ecology of fire, like everything else in a forest, reveals itself a piece at a time, and there is still much to learn about how it works. Fire seems to transform nutrients present in the litter into forms usable by plants, operating faster than normal decay. By sterilizing the soil, fire controls fungal diseases of young Giant Sequoias and keeps root rot and insect infestation in balance. The hottest fires produce the best seed germination and the best seedling survival if followed by moist summers.

On the con side, a burn that occurs in late spring may wipe out spring-flowering plants before they go to seed and hence affect plant succession, at least temporarily. Where no fires have occurred over long intervals, Giant Sequoias sometimes simply outlive the White Fir and other trees that grow up around them and eventually produce a bumper crop of seedlings in the moist decomposition of their fallen competitors' trunks and roots. These seedlings can survive, however, only if the canopy above them is open, letting in sun, and if soil moisture is adequate.

Giant Sequoias have survived for millions of years on their own. The question in thoughtful ecologists' minds, after less than half a century of fire research, remains—what is "natural" for them?

The better known Giant Sequoias alive today are protected in state and national parks. But the majority of Giant Sequoias live in Sequoia National Forest in the southern Sierra under the uncertain policies of the United States Forest Service. For nearly 100 years the Forest Service has protected the Giant Sequoia groves entrusted into its care by the Sierra Forest Reserve Act of 1893.

But between 1982 and 1986 the Forest Service allowed timber companies to log and clear-cut in 5 Giant Sequoia groves within Sequoia National Forest, including Black Mountain. The Service did not remove specimen Giant Sequoias, but it authorized logging of Sequoias less than 8 feet in diameter (2.4 m), along with clear-cutting of their companion trees in the mixed conifer ecosystem. Some Giant Sequoias were left standing like huge isolated sculptures amid stumps and logging debris, subject to the erosion, windthrow, and desiccation of the cutover slopes.

Two local women, Charlene Little and Carla Cloer, stumbled onto such a logged-out Giant Sequoia forest in 1986, in what they had believed were protected groves. The environmental organizations they alerted mediated for a year and a half with the Forest Service, cattlemen, millowners, and recreationists involved.

They emerged in mid-1990 with a compromise settlement hammered out with the help of the Sierra Club Legal Defense Fund. It protects the Giant Sequoias in Sequoia National Forest from logging for up to 10 years, if the Forest Service keeps its word. A presidential proclamation signed in July 1992 extends the mediated settlement indefinitely but has no statutory power. Permanent protection of the Giant Sequoias and their ecosystem can come only through congressional action to include the groves within the boundaries of Sequoia National Park or through congressional or presidential action to declare the area a national monument.

This logging controversy of the late 1980s was a surprise throwback to the days a century earlier, before national and state parks existed, when Giant Sequoias were heavily logged in the southern Sierra.

Despite the fact that the wood of old Giant Sequoias is brittle and breaks into chunks on falling, rendering it useless for lumber, many of the giants were cut down between 1860 and 1890. The durability of Sequoia wood made it useful for roof shakes, grape stakes, fence posts, and flumes to transport water.

Although sometimes called Sierran Redwood for its reddish wood and bark, Giant Sequoia is not the source of redwood decks, paneling, and outdoor furniture. Such items come primarily from the commercially valuable Redwood of California's northern coastal fogbelt.

The enormously wasteful, short-lived Giant Sequoia logging of the late 1800s in the southern Sierra was terminated when a vigorous conservation campaign led by editor George Stewart of the *Visalia Weekly Delta*, Gustav Eisen of the California Academy of Sciences, and John Muir combed the country for support to protect the Giant Sequoias. Writing letters to everyone in and out of Congress who might favor the idea, and to every potential magazine and newspaper advocate, and with the astute help of Robert Underwood Johnson, editor of *The Century*, an eastern United States publication, they caught the attention of Congress.

On September 25, 1890, Congress established Sequoia National Park and surrounding forest preserves. Yosemite's Giant Sequoia groves gained federal protection six days later with Yosemite's designation as a national park on October 1, 1890. The north and south groves at Calaveras Big Trees State Park were added to the California State Park System in 1931 and 1954 respectively, after long, heroic campaigns by the Calaveras Grove Association, the Save-the-Redwoods League, the Sierra Club, the California War Memorial Park Association, and scores of caring individuals, including John D. Rockefeller, Jr.

Giant Sequoias, John Muir's "forest masterpieces," mellow through the centuries, growing ever more imposing with time. Unlike most other trees, they do not seem to die from old age or insect damage. Although over 150 kinds of insects and thirty-seven kinds of spiders and relatives utilize Sequoias in their life cycles, none is known to cause lethal harm, and many furnish food for the birds, shrews, and Deer Mice of the groves.

Old Giant Sequoias die most often by toppling. When growing on the edge of wet meadows, they tend to lean toward the light and to sink into the saturated soil. Growth regulators, called auxins, help to counterbalance the lean by stimulating limbs on the opposite side of the tree to grow faster, bigger, and heavier. But eventually the shallow roots can no longer support the off-center weight and the tree plunges into the meadow. If, in falling, it creates a dam in the meadow, other forest-edge Sequoias may sink into the newly saturated soil and suffer the same fate.

Streams sometimes undercut Giant Sequoias, also causing toppling. Heavy snow loads on the crown can pull over a leaning tree. Carpenter ant galleries in the bark or dead basal wood, combined with a serious off-balance load, may trigger a fall. Root fungus may take a toll.

But the greatest contributor to the eventual collapse of an ancient monarch is fire. Despite the two foot thick, resin-free, and hence nearly inflammable bark on the lower trunks of old-timers, nearly all of them show gaping fire scars at ground level. Even the two largest trees, General Sherman and General Grant, carry huge scars. Ninety percent of Giant Sequoia scars occur on the uphill side where trees grow on slopes, and 90 percent fall uphill, toward their undermined side. Branches and debris rolling downhill through the centuries collect in fuel piles at uphill tree bases and burn into the trees' heartwood, destroying surface roots and weakening the mechanical support.

Even against these odds, Giant Sequoias will often ward off the fall. Their cambium layer, just under the bark, begins to cover a raw fire scar immediately, slowly but surely growing over the wound at the rate of one-half inch a year (13 mm). Stumps show 3-foot-wide scars (.9 m) completely healed in less than 100 years, with the tree growing twice as fast in the injured area as in the rest of the tree.

The Palace Hotel tree in the South Calaveras Natural Preserve and the Room Tree in Sequoia National Park, among others, can both house a dozen people in their fire-hollowed interiors, yet they are very much alive and covering their scars more each year.

A century to heal a wound amounts to a mere blink of time in the lifespan of Giant Sequoias, trees that link the Age of Dinosaurs to the Age of Space.

9

Red Firs and Lodgepole Pines

The best-known Red Fir Forests in California occupy the belt in the western Sierra Nevada between the Mixed Conifer Forests of the midmountain elevations and the Subalpine Forests of the heights. Forming almost pure stands of parklike, mature trees, these largest firs in the world (*Abies magnifica*) reach 180-foot heights (54 m), with rich carmine trunks up to 8 feet wide (2.4 m). At roughly 6,000 to 8,000-foot levels (1,800 to 2,400 m) (higher to the south, lower to the north), they extend from the Kern River drainage of the southern Sierra north into the Cascades of southwestern Oregon, the Klamaths, and the northern Coast Ranges.

California's equivalent of the northern boreal forests of Europe and Asia, the Red Fir Forests find themselves buried each winter under the heaviest snowfall in the state. The storms that blow inland from the Pacific Ocean drop most of their moisture on the 5,500 to 7,500-foot elevations, commonly 13 to 65 feet of snow (4 to 20 m). The area near Tamarack, at 7,200 feet on Ebbetts Pass in the central Sierra, holds the record—73 1/2 feet (22 m) in 1906–1907.

Despite the long, cold, snowy winters, the Red Fir's evergreen needles, lasting 7 to 10 years, enable the tree to produce food at any time when conditions are favorable and to maintain a reservoir of nutrients to draw on as needed. This effectively lengthens the fir's short growing season and provides a competitive edge over deciduous plants that do not burst their buds until late spring or summer. In many parts of their range, Red Firs' new terminal needles emerge whitish gray each

123

Map 9. Red Firs and Lodgepole Pines.

year, giving the trees their common name of "Silvertip." The silvery tips soon turn into blue-green needles that bend upward on branches extending gracefully out from the trunk.

At their lower elevations, where Red Firs grow side by side with White Firs, young trees of the two species look alike, resembling small Christmas trees with silvery trunks full of resin blisters. Mature trees distinguish themselves by the color of their barks, grayish on White Fir, maroon-brown on Red Fir. Old hollow trunks of both are favored choices for bear dens.

When needles are compared, the Red Firs' usually are angled, sharppointed, and bend up at the base; White Firs' generally are flat, blunt, and twist a quarter-turn at the base. But there is considerable variation in all these traits, especially on the upper branches.

Throughout virgin Red Fir Forests, where the floor is often a jumble of fallen trunks, the firs' rosy red bark persists on logs whose interior wood has long since turned to powder. Lack of resin in the wood hastens its decay, but the bark endures.

For many years Red Fir bark was the fuel used in Yosemite's famous Firefall. The bark was heated to red hot embers on the brink of Glacier Point, where the fir grows, 2,000 feet above Yosemite Valley. When the sky was dark and a waiting crowd had assembled on the Valley floor far below, the ranger's call arose from the Valley, "Let the fire fall." And, as eager faces peered upward, the embers were pushed over Glacier Point's granite edge and flowed downward through space like a crimson waterfall. The event drew traffic congestion too heavy for the Valley's limited floor space and was discontinued in 1967.

Red Firs form true climax forests, replacing themselves without a disturbance or a fire. Old trees live up to 500 years. Each autumn they drop seeds from the world's largest fir cones—upright, purplish, resiny cones that disintegrate, scale by scale, on the upper branches. The seeds that germinate the following spring grow into densely clustered miniforests of even age and height. Exceedingly sensitive to the temperature and moisture stresses of high light intensity, the seedlings will thrive only with shade. They reach their peak rates of photosynthesis in very low light levels of about 30 percent. This is not as low as the 9 percent light tolerance of White Fir.

The shade-loving wild flowers of the Red Fir Forests possess the same ability to photosynthesize at low light levels as the tree seedlings. Prince's Pine or Pipsissewa (*Chimaphila umbellata*) and White-leaved Wintergreen (*Pyrola picta*) cram their deep green leaves full of chloroplasts to manufacture as much food as possible in the dim light. Pinedrops, the coralroots, and Snow Plants (*Sarcodes sanguinea*) survive in deepest shade with no green leaves at all. Their secret lies in the fungal network (mycorrhiza) which envelops their roots and feeds them through threadlike extensions that digest the nutrients of the surrounding organic soil.

Pinedrops (*Pterospora andromedea*) produce tiny, ivory-colored, bell-like flowers on 1 to 2 foot high reddish brown stems, sticky with glandular hairs. Some clusters of two dozen or more remain for several years on the forest floor as dried stalks.

The coralroots, which all have widespread distribution in North American coniferous forests, send up stems with small orchid flowers from a root mass resembling coral. The Spotted Coralroot (*Corallorhiza maculata*) grows the length of the Sierra Nevada, abundantly in mountains of southern California, and in the Coast Ranges from the Santa

Figure 28. Red Firs of the higher mountains are home year
round to the vegetarian Porcupine. Hermit Warblers
build nests on upper fir branches in summer.

Cruz Mountains north. Extremely variable in color and size, it appears
in many shades of tan, reddish brown, and yellow, but usually with
purple spots on the flower's whitish lower lip.

The Striped Coralroot (*Corallorhiza striata*) features reddish purple
stripes on off-white flowers, while Western Coralroot (*Corallorhiza
mertensiana*) of the Siskiyou Mountains shows a lower lip either solid
purple or with purple lines.

Snow Plant is almost exclusively Californian, extending only slightly
into southern Oregon and northern Baja California. The brilliant red
plant emerges from snow melt depressions in the late mountain spring
like a fleshy crimson asparagus, its vivid shafts glowing on the dark
forest floor. Red, bell-shaped, 3/4-inch flowers (2 cm) open along the

stems, maturing into scarlet, marblelike berries. In Yosemite, there have been records of thirty-one Snow Plants in some clusters, of individuals up to 2 feet tall, and of others with stems as big around as a man's wrist.

Scattered sparsely under the tall, nearly closed fir canopy, these uncommon flowers share the lower stratum with currants, gooseberries, Bush Chinquapin (*Chrysolepis sempervirens*), and service-berry shrubs (*Amelanchier* spp.). Fallen fir twigs, patterned with white circles where needles have dropped off, cover the forest floor.

Red Fir Forests thrive on shady slopes of deep, rocky soils where there is good soil aeration. Moraines left by glaciers furnish ideal fir habitat. However, in adjacent sunny open glades on gravelly soil, the firs harbor a very different flora. Rosy Pussypaws (*Calyptridium* spp.) hug the ground, spreading spoon-shaped leaves and fuzzy pinkish flowers shaped like their name. Wallflowers (*Erysimum* spp.) and Mules Ears (*Wyethia mollis*) add orange and golden clumps of color. Monardella (*Monardella odoratissima*) enhances its pale lavender and pink hues with a minty fragrance. Nude Buckwheat (*Eriogonum nudum*) shoots up slim, naked stems several feet tall to display small balls of white flowers.

Red Firs share many of their forests with Lodgepole Pines, smaller trees with two short, stout needles per bundle and small prickly cones that grow all over the tree's branches. Lodgepoles prefer "wet feet," and densely populate moist meadows, as well as stream and lake edges. Capable of tremendous control over their internal water balance, they can live in a wide spectrum of soils, but they cannot endure saturation throughout the growing season.

Lodgepole Pines are sometimes mistakenly called Tamarack. Tamarack is actually another name for eastern Larch (*Larix laricina*), a deciduous conifer that does not grow in California.

In some areas Lodgepoles act as pioneers. Their seedlings, which love light, germinate in sunny openings and grow up to create shade for Red Firs. In 100 years or so, the firs outreach and kill the shorter, shade-intolerant pines and take over the forest, usually on slopes above the meadows.

Despite their closeness in living quarters in California mountains, these two higher elevation trees have extremely different geographic ranges. While Red Fir is almost entirely Californian, Lodgepole Pine grows widely over western North America. In the Rocky Mountains of both the United States and Canada, Lodgepole forms dense forests of

Figure 29. Lodgepole Pine forests grow widely over California's high mountain elevations. The trees' twin needles and small cones are often searched carefully for hidden insects by Yellow-rumped Warblers.

trees only 4 inches or so in diameter (10 cm). Periodic fires in the Rockies cause the postfire release of seeds from Lodgepole cones, resulting in thicketlike, small-trunked, even-aged forests. Rocky Mountain Indian tribes once cut these uniform trees into 10 to 15-foot lengths (3 to 5 m), peeled and dried them, and lashed them together as poles to support their circular wigwams; hence the name lodgepole.

California Lodgepole Pines are a different breed from those of the Rocky Mountains, possessing much more individuality, more height, and much bigger trunks—commonly 3 feet in diameter (.9 m), some even fluted at the base. Botanists classify them as *Pinus contorta* ssp. *murrayana*. California Lodgepoles are not fire-type trees. Too little burnable fuel usually grows between them to support a prolonged blaze, and their cones open at normal maturity.

In both the Red Fir belt and in the next higher subalpine zone, California Lodgepoles often replace themselves in continuous pure stands from generation to generation, aided over the centuries by the Lodgepole Needle Miner (*Coleotechnites milleri*). This tiny, gray moth periodically turns Lodgepole acreage into ghost forests by laying eggs in the pine's needles or bark crevices. The emerging larvae kill the needles, and the trees die after three or more defoliations. While acres of dead, silvery trees appear tragic at the height of a needle miner epidemic, all is not lost. The insects' work actually opens up the forest to sunlight in which new Lodgepole seedlings thrive, and the forest gradually replaces itself.

California Lodgepole Pines sporadically occur in moist places and dry ones, all the way from the mixed conifer belt to timberline. They share their Red Fir zone habitat with regular but less abundant Western White Pines (*Pinus monticola*).

Tall impressive trees found over much of the West, Western White Pines thrust out upper limbs that swirl like down-curving wings curled up at the tip. Blue-green, 3-inch-long needles, five to a bundle, clothe the silvery boughs, and clusters of 6 to 10-inch resiny cones dangle from their tips. The barks of old trees crack into striking patterns of deeply fissured squares, somewhat resembling alligator skin.

Although Western White Pines attain their prime growth in the Inland Empire of Idaho and adjacent states, they add a princely touch to the widely spaced, open, sunny higher elevations which they favor in the Sierra Nevada, Cascades, and Warner and Klamath Mountains of California.

Jeffrey Pine, the other prominent conifer of these forests, grows on rocky outposts and boulder-strewn acres throughout the Sierran Red Fir–Lodgepole Pine belt, sometimes joined by Sierra Juniper. The Jeffrey Pines of the drier slopes are often massive trees with stout, reddish brown, plated trunks and big-limbed spreading crowns. The crowns offer nesting holes for Mountain Bluebirds (*Sialia currucoides*) and singing perches for Townsend's Solitaires which pour out a continuous, mellifluous courtship ardor in June. Despite their treetop song posts, the slim gray thrushes choose to nest on or near the ground or in crevices in banks.

Much of the ground is hidden by high-mountain Chaparral, mostly evergreen shrubs that grow especially well around and between the boulders, where reradiated solar heat lasts longer at night. Pinemat Manzanita (*Arctostaphylos nevadensis*) winds red stems over the granite. Huckleberry Oak (*Quercus vaccinifolia*), a true, miniature, acorn-producing oak, forms broad, compact clumps. Sometimes the Washington Lily (*Lilium washingtonianum*) shoots up tall stalks topped with large, fragrant white flowers through the oaks' almost impenetrable midst. If Mule Deer can reach them, they nip off the lily stems as cleanly as a knife cut. Bush Chinquapin and Mountain Whitehorn (*Ceanothus cordulatus*) add cover and seeds for ground-feeding birds of the dry Jeffrey Pine slopes—Green-tailed Towhees (*Chlorura chlorura*), Fox Sparrows, Mountain Quail, and Blue Grouse (*Dendragapus obscurus*).

In autumn, when Mountain Quail migrate downhill, Blue Grouse move into the dense conifers where they survive the winter eating needle tips. With the onset of spring, the male grouse claims a territory of several acres by sending forth a series of deep, reverberant booming notes. He takes a position about 60 feet up in a pine, standing close to the trunk, and for days booms out his call at intervals. Attracted females come to him, and he woos them with aplomb, often choosing several mates over the season. Strutting about with his tail erect and wings drooping, he distends the fiery patches above his eyes. Inflating the air sacs in his throat and neck, he emits a deep, resonant "boom boom boom." The sound carries through the mountain air soft and low, like someone pounding far off.

Among the smaller birds of the Red Fir–Lodgepole Pine Forests, Red-breasted Nuthatches and Mountain Chickadees both eat and cache pine and fir seeds. The aggressive nuthatches use their slender, strong bills to pull insects out of bark crevices as they move up and down tree trunks and probe between the scales of open conifer cones to remove seeds. After hammering excess food into a bark crevice, they sometimes cover it with a piece of lichen or loose bark. Red-breasted Nuthatches depend so heavily on conifer seeds for winter food that in years of poor conifer crops they migrate to other areas.

Mountain Chickadees eat many of the same foods as nuthatches. Heavy insect feeders, they also probe cone scales to reach the seeds of pine and fir and, like nuthatches, sometimes cover extra seeds hidden in bark crevices. Their "dee-dee-dee" notes contrast sharply with the nasal squeaks of the stocky nuthatches when the two species feed together. But, close up, at water holes, nuthatches can be heard making soft, sweet notes.

Chickadees and nuthatches both nest in holes in trees, sometimes adjacent to cavities used by woodpeckers. Normally the birds can all occupy apartments a satisfactory distance apart on the same trunk and

Figure 30. Jeffrey Pines (right) and Sierra Junipers (left) thrive on rocky outposts in the higher mountains, often surrounded by low Huckleberry Oak and Pinemat Manzanita. Blue Grouse (foreground) frequent the habitat.

ignore each other. But occasionally they run into unexpected tenancy problems.

Ward Russell discovered a tree where Mountain Chickadees had chosen a hole one foot above a Williamson's Sapsucker opening, but on the opposite side of the trunk. Ordinarily this would separate the two. But when Russell followed the clamor of young Williamson's Sapsuckers to the nest, he was amazed to see parents of the sapsuckers and the chickadees feeding their young indiscriminately through both nest openings.

The chickadees were carrying insect larvae, the sapsuckers ants. The tiny chickadee parents, taken aback by the violent charge of the much larger young sapsuckers, would retreat to a nearby branch to get up courage to try again. Eventually they managed to deliver the food. In this unusual case, the center of the tree proved to be rotten, and the thin partition between the two cavities had collapsed, dumping all the young together.

Faithful parents, Mountain Chickadees usually remain paired for life and occupy virtually the same breeding territories in successive years. Male and female look alike, tiny gray-backed birds with black bibs, a white cheek patch, and a black cap interrupted by a white line over the eye.

Male and female Williamson's Sapsuckers, at the other extreme, do not remotely resemble each other and were once considered separate species until they were found nesting together. The male is a brilliantly colored bird, possessing a black back, head, and breast, accented with narrow white head stripes, a white shoulder patch, a yellow belly, and a bright red throat. The female bears more conservative tones—a brown head with back, wings, and sides barred brown and white. Both sexes show white rumps in flight, the female sometimes being mistaken for a Northern Flicker.

Like all sapsuckers, this sapsucker drills downward-sloping wells in trees, drinks the sap, ambushes insects drawn to the sap, and eats a variety of fruits and other insects. Its wells have been found in all the conifers of the Red Fir belt, and also in Quaking Aspen. Some of the wells attract migrating hummingbirds, which stake out territories around certain sap holes, sip them every twenty minutes or so, and drive other hummingbirds away from the feast.

Few sapsucker holes escape the all-seeing eye of Steller's Jay. This handsome royal blue jay with a black crest makes itself known at the first rustle of a sandwich bag at all picnic and camp sites throughout the fir and pine forests. An omnivorous feeder, it devours pine nuts, acorns, fruits, seeds, insects, and, on occasion, bird eggs and nestlings. It has been observed seizing juncos and nuthatches in midair and carrying off young Mountain Quail.

Adult Steller's Jays usually reside in an area the year around as permanent mates. They become very secretive at nesting time, choosing a small conifer for their bulky twig nest, using mud as cement.

Young and adult Steller's Jays alike need to stay constantly on the lookout for that skilled avian predator of coniferous forests—the Northern Goshawk. This highly specialized hunter of small birds and mammals dives quickly out of concealed perches to strike with its feet and kill victims with a vicelike grip. Anatomically designed for hunting in wooded areas, the Goshawk's short, broad, rounded wings and long tail provide sudden bursts of speed and the maneuverability to twist and turn and cut between limbs and trunks without breaking a wing.

Lloyd Ingles, watching a Goshawk nest in Sequoia National Park, saw the female "fly swiftly into the thick branches of a nearby fir, about 40 feet (12 m) above the ground. After a few seconds, during which her wings could be heard beating against the small branches, she emerged with a squirming Sierra chipmunk dangling from her talons." Bringing the prey to the nest, "she tore the chipmunk into pieces" for three squealing young birds as they hungrily pecked at her beak (Ingles 1945: 215).

The meal lasted 35 minutes, after which the parent left, and each of the young birds backed to the edge of the platform to forcefully defecate beyond the rim. These same young Goshawks enjoyed meals of Steller's Jays, Northern Flickers, robins, and Golden-mantled Ground Squirrels, plus other local fauna, during their stay in the nest.

Goshawks build nests in several kinds of conifers—Lodgepole Pine, Red Fir, and White Fir in the higher forests, but also in mature Douglas-Fir, Sugar and Ponderosa Pine, and occasionally in aspen. The male generally supplies most of the food and delivers it to the larger, deeper-voiced female, who feeds the young. But she is always on the alert near the nest to pick off a Dark-eyed Junco on the forest floor, a Yellow-rumped Warbler hawking insects, or young Mallards on a pond below. She defends her nest noisily and aggressively against invaders, delivering stunning blows with her feet or raking exposed body parts with her hind claws as she screams a cackling "ca ca ca ca."

Chipmunks and Golden-mantled Ground Squirrels lead busy lives among the logs and boulders of the Red Fir Forests when they can escape the talons of Goshawks and evade weasels and other enemies. Conspicuous small mammals with black and white stripes on their brown backs and sides, they scamper over the forest floor searching for food. The smaller, more delicately built chipmunk has a pointed face with head stripes; the chunkier Golden-mantle sports a plain copper-colored head and neck.

Although both of these spry little rodents use similar holes and burrows under logs and boulders for shelter and nests, they coexist without serious competition because of different food preferences and different ways of obtaining food. Golden-mantles restrict their feeding to leaves and fruits close to the ground. On emerging from hibernation in the spring, they nip off the green shoots and succulent leaves of grasses and herbs, adding manzanita flowers for variety, and continue the leafy diet all summer as new plants emerge. When the plants dry up, they turn to seeds. The Golden-mantles harvest the ground-level nutlets of Mountain Whitethorn while the more agile chipmunks pluck them from the shrub's upper branches.

In addition to seeds, chipmunks favor insects, flowers, and fruits; they are not leaf eaters except in an emergency. Great opportunists, they are quick to spot aphid infestations of fir trees and voraciously devour the aphids for as long as the tiny plant bugs last. On the evenings when winged termites emerge from their nests in logs and take to the air in mating flights, chipmunks appear like magic to feast on them.

During the periodic population explosions of the California Tortoise-shell Butterfly (*Nymphalis californica*), chipmunks and Golden-mantles display their differing predatory skills. In the epidemic years, the caterpillars emerging from eggs laid by this butterfly completely defoliate Mountain Whitethorn and other *Ceanothus* shrubs.

As the spiny yellow and black caterpillars desert the Mountain Whitethorn after chewing up every leaf, they migrate across the ground to nearby firs. En route on the ground, the caterpillars are ambushed by Golden-mantled Ground Squirrels, which take a heavy toll. Chipmunks attack very few of the bristly caterpillars. But after the caterpillars climb the firs and turn into ashy gray chrysalids hanging from the underside of fir boughs, chipmunks scamper nimbly over the branches picking off and crunching the chrysalids nonstop.

Autumn brings a heavy conifer seed crop in good years. This makes up one-third of the Golden-mantle's diet, and while it gathers seeds on the ground, chipmunks garner them in the trees. In poor cone years, both rodents join the other mammals of the forest in digging small pits in the ground, searching for the one remaining available food, the subterranean fungi—truffles. Present in almost every color and consistency, truffles thrive in the soil, attaining peak numbers in the warm autumn when other foods have vanished. Golden-mantled Ground Squirrels, Northern Flying Squirrels, and Douglas Squirrels all apparently detect them by smell.

The lively little Douglas Squirrels, or Chickarees, provide a high proportion of the natural sounds within the Red Fir–Lodgepole Pine Forests, an unbelievable variety of signals. Using mellow mewing calls, loud continuous rattles, buzzes, and scolds, they communicate with neighboring squirrels. The dense tree cover in which they live makes seeing each other impossible much of the time, so, like most of their bird neighbors, they rely on sound to warn intruders and to let rivals know their whereabouts.

Less than half the size of a Western Gray Squirrel, Douglas Squirrels romp up and down trunks and leap nimbly across space from tree to tree. But despite their rampant activity, the tree squirrels are very territorial and live strictly within well-defined boundaries.

Individuals of both sexes set up and vigorously defend their own territories. Averaging from ½ acre to 3 acres, each territory is adjusted to the density of the food available, enabling its owner to harvest and defend a seasonal food supply that will be there all year. Douglas Squirrels are permanent residents in the pine-fir forests, feeding on staples such as conifer seeds and fruiting bodies of fungi, plus seasonal supplements of conifer pollen, berries, insects, and nestling birds. Their brownish gray backs offer excellent camouflage against the barks of most conifers. The buff-white belly separated from the back by a prominent dark line, the bronze feet in summer, the white eye ring, and ear tufts complete a trim, dapper figure.

During the one day of the breeding season when the female's enlarged pink genitals indicate that she is ready to mate, she does not defend her home range, and males from surrounding territories congregate there. Her usual assertive territorial rattle is silent as she listens to the rattles of the competing males.

One dominant male will follow her through the trees and along the ground in a lively, clattering mating chase and mount her when she sits still at intervals. At other times she shrugs him off and runs, apparently teasingly, from tree to ground to tree. Nearby subordinate males watch closely, for if the dominant male loses track of her, they may be lucky. She frequently mates with several males in mountings lasting 1 to 25 minutes. The next day, once again, she is her fiery, independent, vociferous self, defending her territory from all intruders.

When the babies arrive, she provides total care. "The young seem to sprout from knotholes, perfect from the first," John Muir said of them (Muir 1961:186). Fully fledged miniatures of their parents, juveniles are allowed to feed in adult territories for 2 weeks after leaving the nest, then they are driven out to find their own space.

To survive the long, cold winters of the higher forests, Douglas Squirrels build food caches in autumn, many alongside logs or boulders. When the snows come, the squirrels often den under the white cover, next to their cache. This habit makes them vulnerable targets for their prime enemy, the Marten.

The size of a slender, low-slung house cat, the Marten is a tireless hunter of the dense Red Fir Forests. As agile in the trees as any squirrel, and a more powerful jumper, the Marten also follows tracks in the snow to frequent dinners.

As snow melts, it opens cracks of access around the edges of logs or boulders, allowing Martens to plunge into these openings and search beneath the snow. When a Douglas Squirrel and its cache lie at the bottom, the Marten dispatches the squirrel with a bite to the neck, then takes over its den.

Clothed in beautiful brown fur, golden on the throat, this deep forest member of the weasel family is well protected against cold and snow. Active all year, the Marten specializes seasonally on whatever prey is available. During the winter, in addition to digging out squirrels, voles, pocket gophers, and shrews beneath the snow, it stalks Snowshoe Rabbits (*Lepus americanus*) and White-tailed Jackrabbits (*Lepus townsendii*) and ransacks tree cavities for flying squirrels.

In spring and early summer, voles make up its main diet. By midsummer, chipmunks and Golden-mantled Ground Squirrels replace the voles. Autumn finds Martens turning to Mountain Ash (*Sorbus scopulina*) and juniper berries, and insects, mostly yellow jackets.

In the "yellow jacket season," when the black and yellow banded wasps patrol the forests in vast numbers, their sharp tempers and recyclable stingers on the ready, Martens make at least a dent in their colonies. True wasps of a very high social order among insects, Western Yellow Jackets (*Vespula pensylvanica*) live in communities with a caste system of queens, workers, and drones (males). Each colony lasts only one year, from spring to autumn.

The fertile young queens, who are the only yellow jackets to overwinter, emerge from protected hiding places on the first warm days of spring to seek favorable nesting areas. In a hole in the ground, an abandoned rodent burrow, a hollow log, jumbled debris, or occasionally a tree cavity, each queen constructs a small hanging comb of a few hexagonal cells. She lays a single egg in each cell, and when the eggs hatch into white grublike larvae, she hunts ceaselessly to feed them freshly killed flies and caterpillars, while she survives on nectar and honeydew.

This brood emerges into the first workers, which immediately take over enlarging the nest, covering it with a paper envelope and feeding the larvae that hatch from the queen's prolific eggs. Throughout the summer the colony increases steadily to 5,000 or more workers. As the subterranean nest grows, the workers dig out the soil around it to make room for expansion, ferrying loads of dirt some distance away.

They may bring in several thousand insects a day, feeding the queen who has turned into an egg-layer, and feeding the larvae minced pieces, along with regurgitated sweets. The globular nest needs constant additions. Wasp workers, who made paper eons before the Chinese discovered the technique, rasp off shreds of wood from tree trunks, fences, or buildings nearly soundlessly. They gather the wood into little round pellets before carrying it back to the nest. There they work it over with saliva and spread it into thin sheets that dry on the rim of the comb. One yellow jacket researcher estimated that it required 200,000 loads of wood bits to build the light, strong, papier-mâché nest of a colony of 2,000 wasps.

Life within a yellow jacket colony does not always remain peaceful and orderly. Vagabond queens without nests of their own sometimes invade an established colony and challenge the resident queen. The ensuing fight is fierce. The combatants maneuver for a position in which they can use their stingers. When a stinger penetrates one of the few weak spots in the tough armor of the other, the venom injected causes immediate death. If the victorious queen is the invader, she inherits a functioning colony in which to lay her own eggs.

Workers always defend their nests savagely, delivering extremely painful stings if disturbed. The stinging process produces an attractant, an aggregation chemical that is detected by other wasps and brings them to the fray.

As autumn approaches, the workers construct special larger cells in which to produce next year's queens and other cells to produce drones. When these emerge, the virgin queens and drones mate in nuptial flights.

About the same time, the social order of the colony breaks down completely. The old queen dies, worn out from laying 30,000 or more eggs, her wings frayed, often mere stumps from inserting her body into so many cells. The workers consume the remaining larvae in the nest and fly about, feeding on decaying meat, hot dogs, hamburgers, steaks, dropped fruit, whatever they can find in their final days, earning their popular name, "meat bees." The drones die after mating, having performed their sole useful function in the colony. Meanwhile, bears and skunks, along with Martens, dig up the old nests after autumn's cold sets in, relishing any overlooked sluggish larvae.

The new queens seek wintering spots beneath the bark of logs, in wood piles, any sheltered hideout, where they hibernate until spring's thaw brings them out to seek a new nest of their own and start the cycle all over again.

Nearly all birds and mammals of the forests eat insects when they get a chance. The rare, tiny, Flammulated Owl (*Otus flammeolus*), only 6 to 7 inches long (16 cm), feeds on insects almost exclusively, occasionally with tragic results. A Flammulated Owl found dead in a Sierran Lodgepole Pine Forest had tried to swallow an extra large grasshopper. The insect's head had stuck in the owl's throat in a position that prevented either swallowing or disgorging. The bird's stomach contained its other usual smaller fare of crane flies, moths, serpent flies, harvestmen, and caddis flies.

Larger owls prefer larger prey. Primarily a forest dweller by day, the Great Horned Owl often begins its nightly vigil at meadow's edge where voles and pocket gophers offer ready targets. The Great Horned Owl takes rabbits, wood rats, grouse, squirrels, even malodorous skunks and shrews, which the bird, with its poor sense of smell, eats uninhibitedly.

The largest and rarest owl in California, the Great Gray Owl (*Strix nebulosa*), is also a feeder in meadows and glades of the Red Fir–Lodgepole Pine Forests. Hunting mostly at dawn and dusk, the Great Gray Owl perches in stumps or trees overlooking grassy openings crisscrossed with pocket gopher tunnels and mice and shrew trails. Dusky gray all over, with heavy dark streaks, the owl trains its huge, strongly lined facial disks on the ground below. Its riveting yellow eyes watch intently for movement, while its facial feathers, operating like a dish antenna, funnel sound to the ears, pinpointing the source with extraordinary accuracy. On locating prey, the Great Gray Owl swoops down on silent wings, talons outstretched, quickly killing the victim with a bite to the back of the head.

Great Gray Owls do not build nests. They use large snags or broken-off tops of large dead trees, preferring old-growth Red Fir or Mixed Conifer Forest with at least 20 acres of moist, ungrazed meadow nearby for hunting. Sometimes they take over old Goshawk stick nests.

Great Gray Owls are affectionate mates. During courtship the male offers the female a pocket gopher. If she accepts, this seals the bond, and they mate. Then, and at intervals throughout the nesting season, the pair preen and cuddle. Side by side on the branch, they groom each other's facial disks and comb breast and back feathers with their sharp talons. Later, 3 to 4 owlets are born, requiring food for two months before taking their first flights. Although both sexes are expert hunters, it is the male who provides the catch for his mate and for their young.

Great Gray Owls thrive elsewhere in the northern boreal forests of North America, as well as in northern Europe and Asia, but they are on the endangered list in California with numbers estimated at around 50. Most of these live at meadow edges in Yosemite National Park and adjacent Stanislaus National Forest, where their existence depends on the survival of the scarce old-growth forests.

The Great Gray Owl is a close relative of the California Spotted Owl. This Sierran subspecies of the spotted owl, threatened like the northern subspecies with habitat destruction from logging, is distinctly uncommon. Preferring old growth forests but settling for mixed old growth, second growth, and old oaks at times, it lives and hunts strictly among the trees. Flying squirrels, wood rats, and Deer Mice comprise its usual diet. Juveniles in the nest perennially run the risk of detection by Northern Goshawks, Great Horned Owls, and Martens, as well as the Marten's ferocious larger cousin, the Fisher.

Among these relentless predators, only the relatively scarce, powerfully built Fisher owns the equipment to attack and vanquish Porcupines. Porcupines (*Erethizon dorsatum*), protected by their potent quills, are, along with bears and skunks, among the few animals free to amble unconcernedly through the forests. Fishers follow their own random hunting pattern, exploring brushy places, hollow and blown-down trees, and streamsides where squirrels, mice, hares, birds, or chipmunks might be found.

When a Fisher and a Porcupine meet, eyewitness accounts report that the Fisher dashes in, leaps in circles around the slower moving Porcupine to confuse it, while keeping out of range of its adversary's quilled tail. Sharp slashes of the Fisher's needlelike claws and teeth to the Porcupine's face and head quickly kill it. The Fisher then rips open the Porcupine's soft underbelly and finishes off the contents, including most of the bones.

The price of victory is usually a few embedded Porcupine quills which dissolve harmlessly under the Fisher's skin, for the Fisher generally is immune to the quill-caused infections suffered by Coyotes, Mountain Lions, Bobcats, and other Porcupine attackers.

Porcupines in California live primarily in the coniferous forests, where they feed on the tender buds, leaves, and inner bark of trees.

Their taste for inner bark leads to the girdling and death of many conifers, especially saplings in regenerating burned areas. They also kill tree tops by chewing terminal buds and shoots, causing the formation of a disfiguring forked trunk. In the Sierra Nevada they prefer Lodgepole and Jeffrey Pine bark over the firs and often spend the long, cold winters well up a pine, on a branch next to the trunk, feeding on bark and needles amid dense needle clusters.

Some Porcupines move into aspen groves in autumn and survive the winter by eating inner aspen bark. Aspen carries on more photosynthesis in its greenish white bark than in its leaves and supplies necessary winter nutriments, long after its leaves have fallen, for Snowshoe Rabbits, Mountain Beavers, voles, and other mammals.

The most widespread tree in North America, Quaking Aspen in California occupies a narrow zone through the high elevations at 6,000 to 9,000 feet (1,800 to 2,700 m), from the San Bernardino Mountains of southern California to the Oregon border. Its airy, sun-dappled groves enfold clean white trunks supporting leaves that shimmer in the slightest wind on their slender, flattened leaf stalks. Less flammable than the surrounding conifers, aspen groves form natural fire breaks and lend a diversity of habitat that attracts Warbling Vireos and other deciduous-loving birds.

Favoring mountain-meadow and creek-edge soils, aspens spread chiefly through lateral roots that radiate out just beneath the surface of parent trees. An entire stand of aspens usually shares a common root network and identical genes.

On the western slope of the Sierra, all the trees are male clones, formed as suckers from males trees which produce only pollen. This male-only distribution is a puzzle. Elsewhere, aspens do form female trees whose seeds float away amidst fluffy down in the spring winds. Eastern Quaking Aspens have summer rains to germinate their seeds. Western trees have adapted to summer drought, and in many parts of the West, clones prove more effective.

Aspens often act as pioneers for the conifers. Their groves provide shade and shelter in which firs and pines germinate and flourish. The conifers gradually overtop and choke out their benefactors. But aspen roots continue to reach out along forest edges to launch new aspen groves, which, in time, will nourish new young conifers and continue the cycle.

In addition to their pioneering function and the habitat diversity which they provide amid the dark conifers, aspens add a special beauty to these high mountain forests. In autumn, throughout the Red Fir–Lodgepole Pine belt, the golden and coral leaves of white-trunked aspens bring a spectacular last glow to the forest scene before the first snows.

10

Subalpine Forests

Backpackers hiking California's high, scenic trails often feel as close to paradise as they will ever be. Wandering among the rocky forested basins, lush meadows, and sparkling lakes at timberline opens top-of-the-world views. With snow-capped peaks above, wild-flower gardens in the seepages, and marmots sunning on the boulders, life can seem wondrously serene.

These delights on a congenial August day tell only part of the story. Harsh environmental conditions prevail in this zone most of the year. The Subalpine Forests that live at timberline, where trees straggle to their upper limits, climb there through a land ruled by rock and weather. Massive granite slabs, many polished to mirror brilliance by glaciers, hold large erratic boulders stranded eons ago by that same moving ice. Talus slopes of jumbled rock rise to rugged crests. In many areas, the only soil hugs tiny pockets in rock crevices. Fierce winds pick up the shallow, coarse, decomposing granite and blast off the windward sides of trees on exposed ridges. In a winter that may last ten months, temperatures range from roasting to below freezing, with frost possible any night of the year. The only available water is from winter snowmelt, save for an occasional summer thunder shower. Solar radiation is intense.

Not many plants or animals can handle such extreme environmental conditions. The species that succeed have evolved special adaptations to subalpine rigors over time. The ways in which they combat the elements and spread their progeny over this beautiful but relentless landscape make a remarkable story.

California's subalpine domain covers roughly 8,000 to 11,000 foot

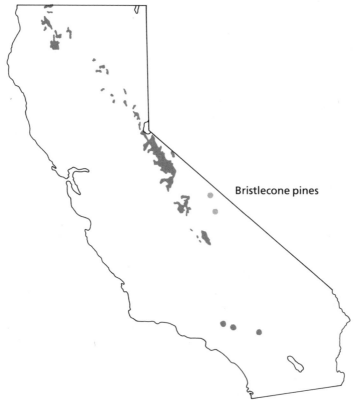

Map 10. Subalpine Forests and Bristlecone Pines.

elevations (2,400 to 3,300 m), with variations in species, as usual, in
different parts of the state. The extensive Subalpine Forests of the Sierra
Nevada, which lie just below the summits along 200 miles of the central
and southern part of the range, will serve as our model of subalpine
ecology, from which to explore briefly subalpine areas of California's
Cascades and Warner and Klamath Mountains. The drier subalpine
heights of southern California's mountains merit separate coverage, as
do the inimitable subalpine Bristlecone Pine Forests of eastern California's White Mountains.

Whitebark Pine and Associates

The Whitebark Pine (*Pinus albicaulis*) is the most characteristic treeline conifer in the Cascades, Warner Mountains, and

Sierra Nevada. In the Cascades, Whitebark Pines share timberline with Mountain Hemlocks (*Tsuga mertensiana*) and Shasta Red Fir on Mount Shasta. They also grow commonly on the high ridges of Lassen Volcanic National Park above a Mountain Hemlock zone and occur in scattered stands on several nearby Cascade summits. The Warner Mountains of northeastern California hold just one peak rising to tree limits, Eagle Peak, at 10,000 feet (3,000 m), which support bands of Whitebarks. In the Klamath Mountains of northwestern California, Whitebark Pines form open woodlands of dwarfed trees above high mountain meadows.

By far, the majority of Whitebark Pines in the state live in the Sierra Nevada. From the Tahoe region south to Mount Whitney, they cover an extensive high mountain realm. Growing as widely spaced trees on protected slopes, they stand 30 to 40 feet tall (9 to 12 m) with rounded crowns, silvery white barks, clustered needles (five to a bundle) at the ends of exceedingly flexible branches. Elsewhere, they form thickets, with multiple trunks sheltering resting hollows of White-tailed Jackrabbits, Mule Deer, and Bighorn Sheep (*Ovis canadensis*). Their knee-high elfinwood mats, often thick enough to hold a heavy hiker, march up to the 12,000-foot level (3,600 m). The species' genetic ability to grow low and horizontally enables it to survive the gales of heights well beyond the reach of its competitors.

Mountain Hemlocks also exhibit the inborn variability essential for success in both the upper and lower reaches of the subalpine zone. In the lower stretches they form almost pure groves in north- or east-facing canyons, where snow lingers well into summer. Tall, narrowly conical trees with soft bluish green needles that look starlike from above, Mountain Hemlocks carry a slightly nodding tip and add a grace to the stately forests where they frequently associate with Red Firs and Lodgepole and Western White Pines. In the northern Sierra, hemlocks are the dominant subalpine species.

But as Mountain Hemlocks ascend the higher rockbound slopes to 1,000 feet past their regular treeline in the central and southern Sierra, they assume the form of sprawling shrubs and gnarled mats close to the ground. The powerful winds that would scour an erect tree trunk bare and knock off its crown, will ride up and over a horizontally growing mat, merely shearing its top. Trees that have evolved this low form of elfinwood (known botanically in German as *krummholz*, for "crooked wood") are the common survivors in mountain heights around the world.

Lodgepole Pines which occupy sunny open slopes and flats in the lower subalpine belt as full forest trees have been slow to evolve the elfinwood mode of survival. As individual Lodgepole Pines climb toward timberline, most of them retain their erect, weathered stance as far up the heights as the elements allow them to grow. However, in a few areas, these pines are showing the genetic ability to form mats.

Sierra Junipers (*Juniperus occidentalis* var. *australis*) survive both ways in subalpine terrain, as mats and as erect trees. Short, thickset, cinnamon-trunked junipers form almost the sole species spaced over miles of granite domes and rocky outposts, their roots delving into seemingly hopeless rock fissures in successful quests for water. In the shelter of large boulders, junipers spread a tough horizontal mat of tight-fitting scaly leaves, angled like a roof to deflect winds up and over. Many of their wind-polished, half-bare, silvery trunks bend to buffer the low green growth in their lee.

Growth is slow for all trees in the austere mountain heights, varying greatly with local exposure, wind abrasion, sun, and the amount of snowmelt. Junipers growing in totally unfavorable microsites have been found to produce 136 annual rings in 2 inches (5 cm) of trunk cross-section. Like fellow slow-growing subalpine trees, they attain ripe old ages of 1,000 to 2,000 years or more. The Bennett Juniper near Sonora Pass, 85 feet tall (26 m) and 14 feet in diameter (4 m), has been estimated at 3,000 years.

Whitebark Pines may live 1,000 years. John Muir counted annual rings on a shrubby Whitebark Pine that was huddled behind a boulder at treeline. The 6-inch-wide trunk (15 cm) was 426 years old.

Just a slight earth rise or hummock makes a world of difference in the microsite of a tree or wild flower. Studies show that an inclined surface facing the sun receives half again more heat and light than a level surface. The importance of this added warmth to seedlings lucky enough to germinate in such favorable spots can be immense. Inclined surfaces also influence the distribution of the lovely White Heather (*Cassiope mertensiana*), the yellow-throated magenta Sierra Primrose (*Primula suffrutescens*), the red Mountain Heather (*Phyllodoce breweri*) and other wild flowers scattered among the rocks and crannies of the subalpine zone.

Spreading Progeny

The conifers of the high rocky slopes disseminate their seeds in a number of different ways. Sierra Junipers produce bluish berrylike cones whose seeds will not germinate unless they have passed through a bird or mammal digestive tract. Townsend's Solitaires, Coyotes, rabbits, and Martens, among others, act as willing agents.

Hemlocks and most pines rely on the wind to disperse their winged seeds, which are kept tightly sealed in cones until maturity. Whitebark Pine seeds are wingless when they leave the cone. Ordinarily this would doom the seeds to drop beneath the mother tree, but Whitebarks have no problem spreading their progeny widely over the high country. The story behind this involves a singular relationship with a bird as much at home in the heights as the pine itself, Clark's Nutcracker (*Nucifraga columbiana*).

A little smaller than a crow, gray with black and white wings and a sturdy, sharp bill, Clark's Nutcrackers begin feeding on Whitebark Pines in late summer, stripping first the closed purple cones and later the dry brown ones of their seeds. They eat some of the seeds on the spot, but the majority fall into a pouch on the floor of the bird's mouth. Roughly 150 Whitebark Pine seeds can fit into the pouch. Only two nutcracker species in the world own such pouches: Clark's in North America and the Eurasian Nutcracker (*Nucifraga caryocatactes*) in Europe and Asia, and both use it in the same way.

With a pouchful of seeds, the bird can still call its harsh "churr" as it flies to one of its specially chosen cache sites, often across ravines or ridges up to 12 miles away. Here it jabs its long, strong bill into the soil to make holes, one at a time, bringing up the pouched seeds and inserting from one to fifteen of them about an inch (2.5 cm) into the hole. It then rakes soil or litter over the burial spot before moving on to the next hole.

Favorite cache areas are south-facing slopes and ridges that are free of snow early in the spring, making the stored seeds accessible as both winter and spring fare. Such sites are scarce, and the birds often join in communal caches.

Nutcrackers live off their cached seeds until the seed crop of the next summer and show remarkable long-term spatial memories in relocating 60 to 90 percent of the caches. Research in the Rocky Mountains and in the Sierra Nevada has indicated that the birds apparently remember certain nearby landmarks such as trees, shrubs, logs, and boulders and triangulate the angles between those and their caches. Even when snow covers some of their cues, they can retrieve seeds from beneath the snow. Nutcrackers belong to the same family as jays and crows, which are considered among the most intelligent birds.

Figure 31. Whitebark Pines provide nuts for Clark's Nutcracker and shelter for the Snowshoe Hare at treeline.

Since nutcrackers feed themselves and their young almost entirely on pine seeds, aside from occasional forays into insects, carrion, bird nests, and juvenile Belding's Ground Squirrels, they stash huge numbers of the nuts. Nutcrackers studied in Wyoming's Rocky Mountains each cached an estimated 98,000 Whitebark Pine seeds one autumn of a prime seed year. Sierran nutcrackers regularly cache up to 30,000 seeds. Such widespread storage means thousands of cache sites for the birds to remember.

It also means that many more seeds are buried than Clark's Nutcrackers need to survive. Some of the unrecovered seeds germinate in the snowmelt of the high country summer. Eventually groves of Whitebark Pines mark the old nutcracker caching sites.

Thus the bird and the tree benefit each other in a mutually useful relationship. The nutcracker transports wingless Whitebark Pine seeds over canyons, ridges, and glacier-carved cirques, unwittingly acting as a colonizer. The pine, in turn, provides a food supply of nuts rich enough in fat and protein to nourish nutcracker adults and young most of the year.

While Clark's Nutcrackers act as the chief dispersal agents for Whitebark Pine seeds, they are "aided" by Deer Mice, chipmunks, and Douglas Squirrels, which all live at subalpine elevations and make numerous large seed caches.

Deer Mice collect seeds from open cones. Their sense of smell is so keen that they probably relocate and eat most of their stores. Chipmunks climb the trees to harvest seeds, and disseminate some of them. Many of their caches are used up during hibernation. Douglas Squirrels nip off Whitebark Pine cones en masse during late summer and store them in heaps at the base of trees. The seeds remain inside the intact cones until eaten and probably only rarely reach ground favorable for germination.

Subalpine Rock Slide Mammals

Other mammals inhabit the subalpine rock slides. The Yellow-bellied Marmot (*Marmota flaviventris*) digs a den under piles of boulders anywhere from the Red Fir and Jeffrey Pine Forests at lower elevations to the rock avalanche areas at timberline. Everywhere it keeps an alert eye on its surroundings from a commanding perch atop a large rock or log, whistling a sharp warning to its fellows if intruders appear. In leisure moments, it spreads its body like a pancake on the granite, soaking up the sun.

The heavy bodied, short-legged marmot, about 2 feet long (.6 m), grizzled brown above and yellowish brown below with a mixture of black and white on the face, is the largest of the commonly visible high country rodents. Strictly a vegetarian, it grows fat during the summer months on the tender leaves and stems of grasses, wild flowers, and shrubs, as well as fruits and berries. But these short trips to minigardens within reach of its den always pose risks; predators such as Coyotes, Mountain Lions, or even rare native Red Foxes (*Vulpes vulpes*) may lurk nearby.

As winter approaches, the marmot retreats to its grass-lined den under the rocks. A complete hibernator, unlike bears and chipmunks, which awaken off and on during the winter, it curls up and sinks into a deep sleep. Its heartbeat drops from 100 beats a minute to four; its body temperature falls from 97°F to 40°F (36°C to 5°C); it breathes about once in 6 minutes and lives on its fat.

In other burrows among the talus slopes. Golden-mantled Ground Squirrels hibernate just as soundly. But the sprightly Pikas or Conies (*Ochotona princeps*), snug in their rockbound dens under the snow, move about during the winter, selecting food from the haypiles of grasses, sedges, wild flowers, pine needles, and shrubs which they collected, sun-dried,and carefully stored during the summer.

When winds blow snow away from their entrances, these little mammals emerge to nibble lichens off the exposed rocks or push through snow to nearby meadows, where they feed on the tips of plants extending above the snowline. Even when 15-foot drifts imprison them within their chambers, they occasionally issue thin, high-pitched calls from below.

Resembling roly-poly bundles of grayish tan fur, with a short head, round ears, and no visible tail, the 7-inch-long Pikas (17 cm) fit their high mountain environs to a tee. With an average body temperature of 104°F (40°C) and thick fur, they thrive on cold and cannot tolerate heat. A dozen species of them exist in the mountains of Asia; two kinds occur in North America, one in Alaska and one in the Sierra Nevada–Cascade chain and the Warner Mountains of northeastern California.

In summer, Pikas share the broken rock piles of California subalpine areas with Yellow-bellied Marmots, Bushy-tailed Wood Rats (*Neotoma cinerea*), chipmunks, Golden-mantled Ground Squirrels, and their chief enemies, weasels and Martens. Very territorial, Pikas mark their own rock boundaries with secretions from eye and cheek glands and reinforce ownership with nasal bleats from rocky lookouts. With every bleat, the whole body jerks forward and the ears twitch upward in an all-out effort. The small face often bears a semi-inquiring expression.

Although squat, the Pika is very agile. It can spring 10 feet while bounding from rock to rock, the bare toe pads on its furry feet providing superb traction. It uses this agility in summer to reach nearby rock gardens where it cuts off choice green plants and carries mouthfuls of them back to sunny drying spots near its den. The plants cure like hay and when the Pika later stores them in airy, well-drained quarters, they retain all winter the natural color and fragrance of well-cured hay.

Also known as Rock Rabbit, the little haymaker hops like a rabbit, chews with a side-to-side motion like a rabbit, has a nose like a rabbit, and is indeed remotely related to rabbits.

The true rabbit of the high elevations, the White-tailed Jackrabbit, is primarily nocturnal and not frequently seen. Very large (one and one-half times the size of the common Black-tailed Jackrabbit of lower elevations), the Whitetail turns completely white in winter and develops "snow shoes" on its feet for foraging on snow. Its droppings under matted Whitebark Pines reveal its favorite daytime resting hollows.

Subalpine Meadow Life

Meadows, large and small, wet and dry, extend among the conifers of the subalpine zone. Tuolumne Meadows at 8,600 feet elevation (2,580 m), ranks as the Sierra Nevada's largest, meandering between Lodgepole Pine Forests over an area roughly 10 by 15 miles (16 x 24 km). Home to grasses, sedges, wild flowers, and small shrubs, Tuolumne and other dry meadows offer fine burrowing terrain for the high country's most obvious open land rodent, Belding's Ground Squirrel (*Spermophilus beldingi*).

Living in groups of 10 to 200, these grayish brown, medium-sized squirrels warn of the approach of intruders with short, piercing whistles of alarm. Sitting erect on their haunches or standing upright like "picket pins," they gauge the threat and decide whether to run for their burrows.

Stretching anywhere from 10 to 54 feet long (3 to 16 m) and lying 1 to 2 feet below the sod (30 to 60 cm), these burrows provide the squirrels' inner sanctum. With at least two surface openings for escape, and with the nesting portions softened by grass, the tunnels offer protection for the young, as well as secluded dens for the 8-month-long hibernation.

To prepare for that hibernation, which consumes most of their year, the squirrels eat nonstop from late spring emergence to early autumn retirement. Chiefly vegetarians, Belding's Ground Squirrels especially favor flower heads, grasses, and seeds. They harvest tall grasses by pulling stems down, a paw at a time, and then lie on their backs clutching their prize.

The squirrels must enter hibernation in September with their early summer body weight doubled and their body fat increased fifteenfold if they are to survive the long subalpine winter. Heavy eating and good fat metabolism are essential adaptations.

The squirrels occasionally add insects, birds' eggs, small mammals, and carrion to their diet. They raid the ground nests of Mountain White-crowned Sparrows (*Zonotrichia leucophrys oriantha*) in willow thickets for both eggs and nestlings. Some Belding's Ground Squirrels practice infanticide, stealing each other's babies. Although both males and females kidnap and kill newborn pink and hairless pups, only males regularly eat them.

Adult and juvenile Belding's Ground Squirrels both face constant danger from predators in their own meadow diggings. Weasels of similar size can easily wriggle through the tunnels. Red-tailed Hawks, Swainson's Hawks (*Buteo swainsoni*), and Clark's Nutcrackers watch for careless juveniles that stray too far from the dens. The transient Swainson's Hawks sometimes utilize a hunting technique of the resident Badgers (*Taxidea taxus*), lying in wait at a burrow's entrance to seize unsuspecting squirrels as they emerge. Badgers, along with Coyotes and bears, dig the victims out of their burrows by night.

Badgers prove particularly formidable enemies. Grizzled gray, low-slung members of the weasel family, they resemble large furry turtles with broad, flat, black heads marked by a dramatic central white stripe and white lateral blazes.

Ambling through Belding's Ground Squirrel colonies, they listen and sniff. On locating their quarry in the chamber below, they dig in with powerful shoulders and clawed forepaws, webbed at the back like a shovel. Excavating swiftly amid a cloud of dust, they close strong jaws on a quivering prey. Adult Badgers need to consume the equivalent of a squirrel a day to keep up their energy.

A transparent inner eyelid helps them see while digging in loose sand. When pursued, they can dig themselves out of sight in slightly over a minute, throwing the soil back with the front feet and kicking it out of the hole with the hind feet in a constant stream.

In poor weather they hole up in their own dens and sleep. But they are not hibernators and hunt off and on through the winter, most commonly digging out soundly sleeping ground squirrels or pocket gophers moving about in their tunnels beneath the snow. Snow does not bother the Badger, who runs over it with a swift gliding motion and can always dig out a warm meal.

Importance of Snow

The inconsistent snowfall in the state's subalpine forest belt from year to year makes a big difference in the lives of much of its flora and fauna. Male Belding's Ground Squirrels usually emerge from hibernation in spring 1 to 2 weeks before the females, tunneling through deep snow to the surface. Because no food is available, they rely on what is left of their stored body fat to keep them alive until snowmelt. When late spring snowstorms hit, many squirrels starve to death.

Mountain White-crowned Sparrows depend on a heavy snowpack to keep breeding meadows wet and green through midsummer. In years of light snowpack, the meadows dry out too soon. June snowstorms also slow the leafing out of willows and aspens, eliminating cover for birds' nests. Conifers bend into a horseshoe or suffer broken limbs under the weight of especially heavy wet snow. Wild flowers that usually blossom in early July may not bloom until mid-August, too late to set seeds.

Dave DeSante's team from Point Reyes Bird Observatory researched for 8 years the breeding landbird community at 10,000 feet elevation (3,000 m) in the Tioga Pass area of the Sierra Nevada. From 1978 to 1985, the team observed nesting birds in prime Whitebark Pine and Lodgepole Pine Subalpine Forests. The 8 years included three with average snowpacks, four with very heavy snowpacks, and one below average.

They found that timing of the snowmelt was crucial and that areas must be free of snow early enough to produce food or birds can't breed. Breeding bird territories often dropped 60 percent in the late summers following long, hard, snowy winters.

Only nine species of birds proved "subalpine regulars," persisting at nests 8 years in a row, not always successfully. The nine hardy nesters were Clark's Nutcracker, Dusky Flycatcher (*Empidonax oberholseri*), Mountain Chickadee, Hermit Thrush, American Robin, Yellow-rumped Warbler, Mountain White-crowned Sparrow, Dark-eyed Junco and Cassin's Finch. Other species that sometimes nest at that height simply moved downhill to snowfree sites when the snowmelt was late.

Acid Snow

During the past decade the *content* of the snowmelt has taken on life-threatening dimensions in subalpine forests and lakes of California and the West.

The millions of tons of sulfur dioxide and nitrogen oxides that automobiles, smelters, powerplants, and various industries shoot into the atmosphere annually have to end up somewhere. In the air these products from the burning of coal, oil, and gasoline react with moisture and are converted into sulfuric and nitric acid. Returning to earth as acid fog, acid mist, acid rain, and acid snow, these pollutants are highly responsible for dire environmental damage.

Acid rain initially drew public concern in the 1970s when research revealed that it was killing lakes in many parts of the world. Canada had lost 10,000 lakes and 40,000 more were endangered. Sweden counted 18,000 lakes sterilized. In the eastern United States, hundreds of lakes in the Adirondack Mountains were dead and several thousand more were dangerously acidic. Many were chemically comparable to vinegar or tomato juice, with no life left except a mat of acid-resistant algae on the bottom.

At that time, two-thirds of United States toxic emissions occurred east of the Mississippi River. Since then, acid rain and snow have been steadily and devastatingly increasing throughout the West.

A report by the National Clean Air Fund in the mid-1980s listed 10,000 western lakes threatened by acid rain. These are lakes distributed through 23 wilderness areas, 55 national forests, and 10 national parks, including Yosemite and Sequoia/Kings Canyon National Parks in California, Yellowstone, Glacier, Rocky Mountain, Grand Teton, Olympic, North Cascades, and Mt. Rainier.

No mountainous area of the West is exempt. The most fragile lakes and forests threatened by acid rain (falling primarily as acid snow) occur near timberline high in the Cascades, Rockies, and Sierra Nevada, far from the sources of the pollutants which are killing them.

In the Rockies, up to half of the emissions come from copper, zinc, and lead smelters in Arizona, New Mexico, Utah, and Nevada; the other major source is coal-burning electric power plants. In California, nitrogen oxides from automobile exhausts make up the chief source of acid precipitation.

Carried by the wind to the high mountains, the contaminants descend with the snow. When the spring melt comes, they pour into the lakes, causing a sudden surge in a lake's acidity. This surge often comes at the precise time when aquatic insects are hatching, fish are spawning, and amphibians' jellylike eggs are in especially sensitive stages of development.

The results are deadly. Even a minor change in a lake's acid content can decimate the plankton that forms the food base for all aquatic life. Frogs, toads, and salamanders, with their highly permeable skins, readily absorb toxic substances in their watery environment. Like canaries in a mine, they unwittingly act as barometers of their habitat's well-being.

A recent survey of 38 high Sierran lakes where native Mountain Yellow-legged Frogs (*Rana muscosa*) once jumped along the banks sent an ominous message: these frogs remained in only one lake.

Scientists monitoring 14 lakes at 11,000 feet elevation on the western slopes of the Colorado Rockies watched a similar story unfold. Once abundant Tiger Salamanders (*Ambystoma tigrinum*) lost two-thirds of their population in 9 years, the drop linked directly to increasing acid snow. Researchers believe that some of the pollutants came from automobiles as far away as Los Angeles.

High Sierran lakes, along with other lakes that lie in granite basins, are especially susceptible to acid snow damage. Granite lakes lack the alkaline deposits that could buffer the acid in the way that baking soda neutralizes acid corrosion on terminals of a car battery. They have no protection at all.

The trees living near these high mountain lakes may pay a heavy price for acid snow exposure. More than 9 million acres of high altitude forests in Germany, primarily fir and spruce, suffered severe acid rain damage in the 1980s from industrial and auto pollution, causing many German foresters to fear that their forests are doomed. *Waldsterben* (forest death) has become an all-too-common word in Germany.

In northeastern United States, dead and dying skeletons of Red

Spruce and Balsam Fir have turned once luxuriant forests into eerie cemeteries, the trees' needles killed by acidic clouds and mist originating in coal-burning plants of the Midwest.

Acidic precipitation soaking into the ground around conifers in California's high mountains is already stunting the growth of some trees by altering the soil chemistry and reducing seedlings' ability to root. Air pollutants are known to induce biological stress in trees, weakening their systems and making them vulnerable to harm from natural stresses such as insects, drought, wind, and frost.

All these deteriorating changes in subalpine life will be magnified a hundredfold if the ozone shield protecting the earth grows any thinner. The ultraviolet light that beams through holes in the ozone layer is especially intense at high altitudes.

Some biologists are convinced that the dramatic plunge in amphibian populations occurring not only in California and the American West but in pristine high mountains from Switzerland to Ecuador and Australia is related to increased ultraviolet radiation. The disappearance of the beautiful little Golden Toad, the subalpine hallmark of world famous Monte Verde Cloud Forest in Costa Rica, has alarmed everyone.

As herpetologist David Wake put it, "Amphibians have been around for over 100 million years. They're survivors. They survived whatever knocked out the dinosaurs. So, if they're beginning to check out, we'd better take it seriously. The same water that no longer supports amphibian populations in the Sierra Nevada is the water we drink in Berkeley and San Francisco. It seems to me there's a message here someplace. We'd better figure out what it is" (Wake 1990:A5).

Worldwide long-term studies now underway are probing for more details and answers. Meanwhile, if future life on this planet is to continue, our earth must reduce and recycle its pollutants and rid itself of the chlorofluorocarbons and related chemicals in refrigerators, air conditioners, and other appliances that poke holes in the ozone layer. High mountain lakes and forests are sending a warning to those with ears to hear.

Foxtail Pines

Several other distinctive subalpine conifers occupy limited disjunct ranges in California's high mountains. In the southern

Figure 32. Foxtail Pines of timberline get their name from
dense clusters of needles resembling a fox's tail.

Sierra Nevada, Foxtail Pines (*Pinus balfouriana*) thrive on the most
inhospitable rocky ridges and slopes near timberline. Forming pure
stands of widely spaced individuals beneath craggy peaks, they typically
face the savage climate of the 10,000 to 12,000 foot heights (3,000 to
3,600 m) with erect, single trunks, broad at the base, tapering to a dead,
bleached tip 40 feet up (12 m). Cinnamon bark, broken into squarish
plates, covers boles that slowly widen over 1,000 to 3,000 years.

Longevity is also built into the short needles that clothe branch ends
with dense clusters resembling a fox's tail. Like many subalpine conifers, Foxtail Pines conserve energy by retaining needles much longer
than pines at lower elevations, often up to 17 years.

Like other pines of the world that can tolerate extreme cold, Foxtail
Pines survive the severe subzero winter temperatures of the glacial basins by draining the water from their living cells into the intercellular
spaces. Only pines with this inherited capacity to "drain their pipes"
can withstand the bitter cold.

While the erect, often battered trunk is the Foxtail Pine trademark,
occasional trees divide into multiple trunks, and some have been discovered above upper Monarch Lake in Mineral King creeping low in
the squat, bushy forms of krummholz. These krummholz pines may
also be found on the ridge north of Rocky Basin Lakes, south of Sibe-
rian Meadow. Although standing apart from other species is their

norm, Foxtail Pines sometimes join mixed forests of subalpine conifers; at treeline on Mount Whitney, they mingle with Lodgepole Pines.

Three hundred miles to the north of the high Sierran Foxtail pine forests, a separate population of Foxtail Pines thrives in scattered stands in the Klamath Mountains, Trinity Alps, and Yolla Bolly region. The two widely separated populations are believed to be remnants of an extensive Subalpine Forest that existed in California 12 million years ago when summer rainfall prevailed. As the climate changed, only the Foxtail Pines in the state's northwest corner and in the southern high Sierra survived.

Southern California's Subalpine Forests

Southern California's subalpine zone, much drier than that of the Sierra, tends to starve and dwarf many of its conifers, especially on arid, south-facing slopes. But Limber Pine, equipped to handle steep, dry, rocky sites low in nutrients, meets the environmental challenges with ease.

Basically a Rocky Mountain species, Limber Pine (*Pinus flexilis*) extends spottily along the high Sierra from Yosemite south, primarily on the thirsty east side. In the infrequent places where Limber Pine shares terrain with Whitebark Pine, as in Inyo County's Onion Valley, the two species look almost identical. Only the cones distinctly separate them: Limber Pine's huskier brown cones drop intact; Whitebark Pine's smaller purple cones disintegrate on the tree, leaving spindlelike cores attached to the upper branches. Ordinarily, Whitebark Pines grow in snowy timberline zones north of Mount Whitney, whereas Limber Pines occupy dry, windy sites farther south and east.

On southern California's high peaks, Limber Pine forms major timberline forests. The pine's growth habits are unique. The trunk is apt to be crooked and short, with lower branches almost as big as the main trunk dipping toward the ground. Trees only 18 feet high (5 m) may have trunks 2 feet in diameter (60 cm). Young twigs are so flexible that they can often be tied in a knot without breaking. The cones carry large, heavy, wingless nuts, which Clark's Nutcrackers harvest and cache in the same manner as with Whitebark Pines to the north.

Limber Pine dominates the upper Subalpine Forests of southern California's desert-facing mountains. From the Santa Rosa and San

Jacinto Mountains through the San Bernardinos and San Gabriels to Mount Pinos, Limber Pines form the prominent highest altitude forests, generally intermixed with Lodgepole and Jeffrey Pines and White Fir. On Mount San Gorgonio, Limber Pine grows as prostrate krummholz near the summit, 11,700 feet (3,508 m).

Lodgepole Pines also form extensive forests on the higher southern California mountains, sometimes growing clear to the top. On Mount San Antonio (Mt. Baldy), Lodgepole Pines climb out of an open, lower, stunted forest where they mix with Jeffrey Pine, White Fir, juniper, Greenleaf Manzanita, and Mountain Whitethorn to become relatively dominant in the upper forest and then form a low, matted, twisted krummholz near the summit, 10,234 feet (3,070 m).

On some north-facing slopes, Lodgepole Forests shelter a special kind of subalpine meadow, the snowmelt gully, which melts slowly enough to allow seepage during a long growing season and produces a rich meadow flora, primarily Sierran. Grasses, sedges, Shooting Stars (*Dodecatheon redolens*), White-flowered Bog-orchids (*Platanthera leucostachys*) (formerly Rein Orchids), Corn Lilies (*Veratrum californicum*), and Lemon Lily (*Lilium parryi*) are part of the luxuriant parade.

On Mount Baden-Powell, a particularly handsome Lodgepole Pine forest climbs the north slope to join a Limber Pine forest at the 8,634-foot level (2,590 m). The Limber Pines, scattered at first, gradually become an extensive stand of gnarled, ancient-looking trees as they take over dominance clear to the summit at an elevation of 9,558 feet (2,867 m).

The understory of the Limber Pine forest includes tough, drought-resistant shrubs: tall, round-headed Curl-leaf Mountain Mahogany (*Cercocarpus ledifolius*) with its masses of dramatic fruits, radiant in backlighting; Rubber Rabbitbrush with silvery green foliage augmented in late summer by brilliant golden flowers; and Bush Chinquapin with sturdy evergreen leaves and burry fruits. On the ground the pincushion buckwheats, penstemons, mints, Senecios, and other flowers adapt to the arid heights.

Bristlecone Pine Forests

Two miles above sea level in the White Mountains of eastern California grow the world's oldest known living trees—Western Bristlecone Pines (*Pinus longaeva*).

Though no more than 30 feet tall (9 m), bristlecone pines have lived in remarkable harmony with their harsh environment for thousands of years. Relatively unnoticed until 1958, when Edmund Schulman announced the discovery of specimens more than 4,000 years old, they have proven amazing in many ways.

Living under the most unsparing mountain conditions, 9,500 to 11,500 feet high (2,850 to 3,450 m), in an area of intense sun, ravaging winds, thin dolomitic soil and only 12 inches (30 cm) of precipitation a year (mostly snow), bristlecone pines hang tenaciously onto life. Growing extremely slowly, they add an inch or less (2 cm) to their diameter every century. Their short curved needles, five to a cluster, live up to 30 years, reducing the need to produce new ones annually, while providing a stable food source.

When a part of the tree's crown is killed by winds, lightning or drought, an equivalent amount of bark and water-conducting xylem tissue dies back so that the tree keeps its food and water supply in balance. What remains is healthy. All bristlecone pines over 1,500 years old, sandblasted 95 percent bare, have only 10 inches or less (25 cm) of narrow bark lifeline running up a trunk.

The beautiful wood of some of these old barkless trunks is one of the major attractions of the Schulman Grove. Its Discovery Trail passes dozens of ghostlike trees with silvery white branches stark against the deep blue sky, cinnamon-colored trunks with curling grain, soft beige trunks with black grain, and swirls of rich red-orange wood.

Not all bristlecone pines live so incredibly long, to become 40 centuries old. The trees that form fairly dense forests on the moister north-facing slopes in the White Mountains grow straight and even, usually dying of heart rot in 1,000 to 1,500 years.

It is the oddly shaped trees living on the driest, toughest sites that live the longest. The narrow ribbons of bark on their trunks forcibly slow their growth, resulting in dense cells and abundant resin. This dense, resinous wood seems to be what supports life for millennia, as in the case of the 4,700-year-old tree named Methuselah. Even after death, a bristlecone trunk may not fall for an additional 2,000 years. And when it does topple, its wood on the ground may endure another 4,000 years.

Bristlecone forests reveal at a glance the kind of soil underfoot. The bristlecone pines scatter in groves on the pinkish white dolomite, unbothered by its high reflectivity and alkalinity or its low supply of potassium and phosphorus. Companion Limber Pines and a sagebrush understory grow mainly on adjacent granite, some Limber Pines attain-

Figure 33. Western Bristlecone Pines, the world's oldest trees,
grow in shapes dictated by the harsh environment of
mountain heights. Golden-mantled Ground Squirrels
scamper among down trunks.

ing 1,600 years of age. Bristlecone and Limber Pines look somewhat alike in their twisting forms. Both are five-needle pines, but Bristlecone Pine needles grow in a stiff, distinctly bottlebrush pattern and their cones form sharp bristles. Limber Pine needles form dense tufts at the ends of short branches and their cones lack bristles.

The Western Bristlecone Pine forests in the White Mountains are the most westerly extension of these primarily Great Basin trees found over high arid mountains in six western states. As in all of California's Subalpine Forests, they consist of the hardiest survivors.

Each conifer anywhere in the state's high country adapts in its own way to the strong, desiccating winds, the prolonged cold, the short growing season, the sterile, shallow soils, the scarce water, and the sheltering boulders of mountain heights. Each depends on either the wind or on birds and mammals to distribute its seeds. Each tends to live an uncommonly long life, improving the odds for producing offspring that will be lucky enough to germinate and grow in one of the few favorable spots.

Western Bristlecone Pines, ruggedly sculptured, richly colored, a notch above in endurance, unquestionably add the "crème de la crème" to California's Subalpine Forest domain.

11

Pinyon Pine–
Juniper Woodland

Winds carrying Pacific Ocean moisture into California drop most of it on the Coast Ranges and the western slopes of the Sierra Nevada-Cascade chain, and the mountains of southern California. Very little rain and snow make it over the high peaks to the state's east side. The arid lower eastern slopes and their valleys consequently really belong to the Great Basin desert that stretches inland a thousand miles across Nevada, Utah, Colorado, and Wyoming to the Rocky Mountains.

This high desert, ranging from 4,000 to 6,000 feet (1,200 to 1,800 m) at its Sierran border, is an open, clear-air land, colored the pale silvery green of shrubby sagebrush over miles of rolling terrain. Stubby, dark forms of sturdy little conifers mark a belt of greater moisture on the adjacent higher foothills. These are the trees of the Pinyon Pine–Juniper Woodland.

Consisting of junipers and Singleleaf Pinyon Pine, also called Piñon Pine (*Pinus monophylla*), the woodland trees emerge at spaced intervals out of a shrub and herbaceous understory. Classed as a woodland because of the short, widely scattered trees, the junipers and Singleleaf Pinyon Pines form a zone between the denser, taller coniferous forests of the higher mountains above and the sagebrush flats below.

The woodland trees require 12 to 20 inches (30 to 50 cm) of precipitation a year, whereas sagebrush and its companion shrubs can survive on 7 inches (17 cm). Thus, over much of California's eastern high desert, sagebrush communities in the drier areas alternate with mixed woodland-sagebrush in the wetter places.

162

Map 11. Pinyon Pine–Juniper Woodland.

Water is the key to plant distribution in the Great Basin. Available moisture ranks as the single most critical environmental determiner in this western region. Every ecosystem weaves its web of life around the precious fluid.

The Pinyon–Juniper Woodland ecosystem includes many inter-dependent plant and animal relationships, from Sage Grouse to Pinyon Jays to kangaroo rats, all linked to food and water. The dominant pin-yon pines and junipers use water very sparingly, conserve it within their tightly knit youthful frames, and grow so slowly that it often takes 100 years to attain a 6-inch-wide (15 cm) trunk.

Three species of pinyon pine occur in California. Singleleaf Pinyon Pine is by far the most common, ranging along the eastern Sierran slopes from Sierra County south, with forays into the Tehachapi Moun-tains and sporadic areas west of the crest. In southern California the Singleleaf Pinyon Pine is usually replaced by the four-needled Parry Pinyon Pine (*Pinus quadrifolia*) which grows amidst Chaparral, not open woodlands, in scattered locales from the Santa Rosa Mountains to Baja California. The two-needled Colorado Pinyon Pine (*Pinus edulis*), the chief pinyon of the central Great Basin and of Grand Canyon, Zion, Canyonlands, and Mesa Verde National Parks, appears in California only in two desert ranges of San Bernardino County.

Singleleaf Pinyon Pines are the only pines possessing single round needles. All of the other 100 kinds of pines in the world bear 2 to 5 needles in a cluster. Short and sharp-pointed, the needles cover the squatty young pinyons with a bluish gray foliage that contrasts with the yellow-green of their juniper associates. As the pinyons get older, they grow taller, lose the rounded or beehive shape of youth, branch more widely, and flatten out on top, providing improved shade for Mule Deer, chipmunks, ground squirrels, wood rats and other grove wildlife. The trees remain small and widely spaced on drier sites near desert margins, while on higher, moister, more favorable locations, they form dense, pure stands 30 to 40 feet (9 to 12 m) in height.

The Singleleaf Pinyon Pine is best known for its highly valued pine nuts. The Washo and Paiute tribes found California's pinyon woodlands to be an Eden. The annual pinyon pine nut festival and harvest gave focus to their year.

Margaret Wheat recounts the Paiute harvest. Each fall when "the rose hips turned red in the valley, then they knew that the pine nuts were ripe in the hills." Backpacking their possessions, whole families worked their way up the slopes to the pine groves, where they set up camp. Starting at dawn, men walked back and forth under the trees, beating the branches with long hooked willow poles, showering cones, nuts, pitch, and twigs on the women and children gleaning nuts from the ground. They worked carefully, as there was a taboo against breaking the limbs of their provider.

Young boys scouted out more good trees, while "girls hauled the loaded baskets back to camp where their grandmothers and aunts sat together cleaning the nuts for roasting" (Wheat 1967:13, 31). Individual Owens Valley Paiutes often gathered up to forty bushels in a season.

At midday they all napped. Evenings around the bubbling, pitchy pinyon pine campfire, amid the fragrance of roasting nuts, the elders told the children stories of how their ancestors first got the pine nuts.

Pine nuts, a highly nutritious food, formed a major part of Native Americans' diet throughout the Great Basin. Roasted, raw, or made into soup, the nuts possess all twenty of the amino acids and contain the same percentage of protein as a pecan. They have less than half the fat (85 percent unsaturated), and five times more carbohydrate. Peanuts are higher than pine nuts in both protein and fat and lower in carbohydrate.

In the past, woodland mammals and birds competed with Indians for the nuts. Today, in the groves, they still vie with each other for this favorite food. Large-eared, white-footed Pinyon Mice (*Peromyscus truei*) climb to the treetops for it. Chipmunks, Golden-mantled Ground Squirrels, and wood rats claim their quota. White-breasted Nuthatches, Clark's Nutcrackers, chickadees, and other birds of the adjacent Sierran coniferous forests all take advantage of the handy food source.

But the bird linked most closely to the pinyon pine is the Pinyon Jay (*Gymnorhinus cyanocephalus*). This jay of dull bluish gray color thrives in pinyon pine regions, associating with the trees in ways that benefit both species.

Gregarious birds, Pinyon Jays live in flocks of 200 to 300, wandering widely when food is scarce but sticking to a home range in Pinyon Pine Woodlands when the pine nuts are available. In the years when pinyon seeds are plentiful in late summer and fall, the jays harvest them in vast numbers, eating some and transporting more to communal caching areas in their nesting groves.

Possessing an expandable esophagus that holds up to fifty-six seeds, each jay can transport enormous quantities of food over many visits. They choose spots on the south side of pinyon pines or alongside logs or rocks where snow melts early in the spring, scraping the litter from the ground, inserting fifteen or more seeds in the soil, and covering them with the litter.

When breeding gets under way, the birds utilize the stored nuts as their chief food source. Forming permanent pair bonds, Pinyon Jays are with their mates throughout the year and can begin breeding whenever environmental conditions are favorable, sometimes as early as February.

The birds live in loose colonies, with only one nest per tree, usually 8 to 10 feet up (3 m) in a juniper or pine. Well insulated and often facing south, the nests receive almost steady attention from the female as she incubates her eggs, preventing them, and later nestlings, from freezing in the cold Great Basin winter. Her mate leaves the foraging flocks to bring pinyon pine seeds and insects back to the nest. During their final days in the nest, all young are fed communally.

After the young jays leave the nest but are still undeveloped fliers,

Figure 34. The Singleleaf Pinyon Pine, the only pine with single, round needles, shares a close relationship with the Pinyon Jay.

they become members of group nurseries. Bands of twenty-five to fifty fledglings stay close to home in trees watched over by adult "baby-sitters." Other adults bring back food to feed them all.

This is the breeding pattern of Pinyon Jays when Singleleaf Pinyon Pines produce bumper autumn crops. In years when there is no fall crop, the jays do not breed until nuts are once again available or until wet weather brings abundant insects for their young.

The benefits of this unique interrelationship to both the jay and the pinyon pine are obviously immense. The jay secures highly nutritious food, shelter, and nesting sites. The pine acquires an efficient seed dispersal service. Singleleaf Pinyon Pines produce large wingless seeds. Without a means of spreading them, their nuts would fall to earth, hit or miss, beneath the parent tree—on rock, in the hot sun, with slim chance for survival. The jays transport the seeds to other areas and by caching them, effectively plant the seeds in soil, preventing surface drying. The number of seeds planted provides favorable odds for successful germination of the hardiest.

Furthermore, the jays screen the nuts very carefully on the tree before selecting them. Pinyon pine cones open wide, displaying the nuts while holding them in place with a thin membrane. The nuts come in two colors: dark chocolate brown shells which contain firm, tasty meat, and tan shells which are generally light and empty. Pinyon Jays choose only the dark brown seeds and then weigh the seeds briefly in their beaks before clicking the bills rapidly on them. A bad seed rings hollow. Only seeds that pass the tests of color, weight, and sound get planted.

In the northern part of their wide Great Basin range, where pinyon pines are absent, Pinyon Jays live on juniper berries and seeds of other pines, so they are not tied to the pinyon for survival. But where both occur together, the relationship is close and mutually beneficial.

Junipers

Three species of shrub or tree-size junipers live along California's east side—Western Juniper, Utah Juniper, and California Juniper. North of Lake Tahoe, Western Junipers (*Juniperus occidentalis* var. *occidentalis*) form a Northern Juniper Woodland dominating the lava flows of the Modoc Plateau, continuing into some interior localities, and up through eastern Oregon. In certain areas they fill in the transition zone between the mixed conifers of the forested slopes and the lower sagebrush flats.

Tough, sturdy trees, these Western junipers are built to endure the high desert's bitter winters and hot summers. Growing in all shapes and sizes, they take on an unkempt look when the pinkish outer bark layer curls and shreds, exposing reddish brown underbark. The tightly overlapping scalelike leaves reduce water loss and show numerous tiny, sticky white dots that glisten like diamonds in the sun.

The Devil's Garden near Alturas in northeastern California hosts a large stand of mature Western Junipers. On early summer mornings, the Garden's sagebrush–Antelope Bush (*Purshia tridentata*) shrub understory is annually enlivened by a colorful wild flower display.

Whole fields of white yampah (*Perideridia* spp.) and bright yellow composites share the rough terrain with ivory, red, and yellow buckwheats and blue penstemons. Silvery grass tufts soften the volcanic red and black rocks. The three-lobed white ray flowers of foot-high Blepharipappus (*Blepharipappus scaber*) wave ghostlike between juniper logs festooned with chartreuse Staghorn or Wolf Lichen (*Letharia vulpina*).

As everywhere in the West, the original human inhabitants of Northern Juniper Woodlands found ways to incorporate local plants into their lives. Staghorn Lichen was boiled in water by Klamath Indians to produce a bright canary yellow dye for their baskets. Yampah's starchy tuberous roots were, and are, a savory staple, eaten fresh or cooked or ground into flour for baking.

Most of the plants harbor insects or seeds sought by the birds. Scrub Jays screech as they search juniper boughs. Mourning Doves peck the soil between rocks for tidbits and take off on whistling wings when alarmed. Northern Flickers' staccato calls ring loudly through the open woodland. Ash-throated Flycatchers chase a brilliant blue male Mountain Bluebird from a potential nesting hole.

Plain Titmice parents feed their begging, wing-fluttering young in lower juniper branches between unceasing hunts for more food among the scaley leaves. Buzzy trills on pockmarked boulders pinpoint two small tan and white Rock Wrens dipping and bobbing from rock to rock.

The Northern Juniper Woodland attracts other bird species in the snowy winters. Townsend's Solitaires spend the cold season here, moving down from summers in the high mountain forests. Contrary to most birds, which defend territories around their nest only in the breeding season, solitaires establish winter feeding territories around their favorite juniper berry trees and defend them from other solitaires.

When the snow cover is thin and exposes berries on the ground, the slim gray thrushes forage there. When the ground lies under snow, the solitaires hover to pick off berries from branch ends. Robins, Steller's Jays, Western Bluebirds, and Scrub Jays also eat juniper berries and are winter competitors, along with Coyotes, Deer Mice, and Martens. Western Bluebirds show a special fondness for berries of the mistletoe that attaches itself to many junipers and sometimes roost at night in mistletoe clumps.

South of Lake Tahoe the smaller Utah Juniper (*Juniperus osteosperma*) joins Singleleaf Pinyon Pines in the Pinyon–Juniper Woodland that runs the length of the Sierra Nevada and east into the Great Basin. In some southern California areas, such as the San Bernardino Mountains, Utah Juniper fills in the middle belt at 4,500 to 6,000 feet (1,350 to 1,800 m) between its relatives, the shrubby California Juniper (*Juniperus californica*) of the lower desert elevations and the Sierra Juniper higher up (*Juniperus occidentalis* var. *australis*). Utah Juniper often forms pure stands below the levels where it merges with Pinyon–Juniper Woodland. It pioneers the formation of new woodland by creating the shade in which the pines can germinate.

Able to thrive in the most drought-ridden spots, Utah Juniper conserves available water through its coarse greenish foliage and stiff branchlets. Its half-inch-wide berries rank among the largest of the junipers. Actually fleshy cones with seeds inside, the berries turn bronze when mature. Juniper berries from various parts of the world provide the flavoring for gin. The crushed foliage of Utah Juniper yields a pungent, ginlike aroma.

Peaggies

The Jeffrey Pines that grow regularly in the next higher mountain zone above the Pinyon Pine–Juniper Woodland harbor numerous insects. Among them is one which the Paiutes of the Mono Lake and Owens Valley regions traditionally added to their food supply.

Each July the tribes traveled uphill to the Jeffrey Pine forests at a time when the larvae of the Pandora Moth (*Coloradia pandora*) were descending from the tree crowns, after literally defoliating the pines. Known to the Paiutes as "peaggies," these 3-inch-long greenish brown larvae were choice morsels. To capture them, the Paiutes dug shallow trenches around infested trees and set smudge fires around the tree bases. The smoke forced the caterpillars to drop to earth in vast numbers. Scooping them out of the trenches, the Paiutes cooked and dried them and mixed them with vegetables in a stew.

Some tribes smoked the peaggies in heated dirt mounds. Klamath and Modoc Indians of northeastern California preferred waiting until the descending caterpillars buried themselves in the trenches and turned into pupae, when they collected and roasted them. The pupae form the overwintering stage of the 2-year cycle.

The adult Pandora Moth differs from most moths in being active during the day. Large, heavy-bodied, and grayish brown, with one black spot in the center of each wing, it has a 4-inch wing spread (10 cm). A devastating defoliator of pines from the Rockies west, it commonly lays eggs on Ponderosa and Lodgepole Pines, in addition to Jeffrey Pines.

About every 20 to 30 years, it balloons to epidemic proportions, stripping vast acreages of pines bare. East side Indians in California were well aware of the moth's normal 2-year cycle and counted on their alternate-year cornucopia of peaggies.

Shrub Understory

The Black-tailed Jackrabbits, Coyotes, rodents and other berry, seed, and leaf eaters of the pinyon-juniper community spend most of their lives among the woodland's shrub understory. Big Sagebrush (*Artemisia tridentata* ssp. *tridentata*), Antelope Bush, rabbitbrush, Green Ephedra (*Ephedra viridis*), and other shrubs provide the cover for diverse populations of insects, lizards, snakes, and small mammals, as well as hunting, resting, courting, nesting, and sleeping space for all the wildlife.

Big Sagebrush, probably the most abundant shrub in the West, dominates the shrubland. Preferring the rainfall of the foothills, but able to survive on less, it sinks its stout taproot 4 to 10 feet (1 to 3 m) into deep soils. It can reach heights of 10 feet (3 m) under prime conditions and can grow so densely in places that Mule Deer have to thread a path between the tannish gray shredded stems and fallen silvery branches.

The chemical compounds in sagebrush which fill the air with such delightful pungency after a rain and produce aromatic smoke from a sage campfire are largely unappetizing to cattle. However, the leaves are eaten by deer, Pronghorn (*Antilocapra americana*), jackrabbits, and ground squirrels.

Dull gray in flat midday light, the many species of sagebrush come luminously alive in the long slanting rays of early morning and late afternoon. These are the hours when sagebrush birds make themselves known. The Sage Sparrow (*Amphispiza belli*), gray with black whiskers and a single breast spot, forages among the bushes and sings from their tops. The pale brown Brewer's Sparrow (*Spizella breweri*) pours out a sweet bubbly mixture of trills varied in rhythm and pitch. Green-tailed Towhees often dominate the avian chorus, showing off their iridescent green wings and tail and rufous cap as they combine clear fluid notes with burry ones and catlike mews.

But the bird that is 100 percent linked to sagebrush, and occurs nowhere else, is the Sage Grouse (*Centrocercus urophasianus*). America's largest grouse gets everything it needs from sagebrush—shelter, food, nesting cover, mating arenas. Feeding almost exclusively on sagebrush leaves, it lives just a pecking distance away from a satisfying menu the year round.

At daybreak in early spring Sage Grouse gather on their ancestral strutting arenas, known as leks, to reenact one of the most bizarre spectacles in the West. The grayish-brown-backed males (cocks), about the size of small turkeys, stretch into upright strutting posture, heads held high, chest sacs sagging, wings drooping, tails erect in a spikelike fan.

As the cocks step forward in their strut, they raise and lower their white chest sacs, filling them with air and exposing two yellow bare skin patches resembling "sunny side up" fried eggs. When the bare skin patches quickly collapse, the cave-in is accompanied by sharp snaps and a soft cooing sound. Meanwhile, the wings swish forward and back.

The cumulative effect of these sounds and actions is a mesmerizing chorus of booms, clucks, plops, coos, and swishes as the cocks parade to outbluff each other in the male hierarchy and gain dominance as the master cock. The master cock's territory includes the highly prized mating center where the smaller, quieter females copulate with the master cock. After mating, the females fly off to build nests and raise young on their own.

Nearly as abundant as Big Sagebrush, and sharing with it an altitudinal range from lower sage land to timberline, Antelope Bush is a shrub much more palatable to wildlife of the pinyon-juniper belt than is sagebrush. Growing up to the same 10-foot (3 m) heights as Big Sagebrush in favorable sites, Antelope Bush sends down roots to 15 feet (5 m). At a distance the two shrubs are not easily distinguishable, but close up, the narrow darker green three-lobed leaves of Antelope Bush separate it from Big Sagebrush's wider, more whitish, three-lobed leaves.

Although both are beautifully adapted to survive, and thrive side by side in the aridity of the Great Basin, they belong to different plant families. Antelope Bush shows its affiliation with the rose family in June when it covers itself with creamy blossoms. Sagebrush blooms from August into October, producing clusters of inconspicuous tiny, yellowish green flowers grouped in a head, characteristic of the sunflower family.

The heavy seeds of Antelope Bush are popular with birds and rodents. Ground squirrels, chipmunks, Deer Mice, pocket mice, kangaroo rats, and wood rats all relish them and sometimes spread the shrub by caching its seeds. Antelope Bush leaves and young twigs are a favor-

Figure 35. Bushy-tailed Wood Rats build stick nests on the
ground or beneath rock outcrops in most California
forests. They furnish juicy morsels for Coyotes,
Bobcats, and large owls.

ite winter food of the Mule Deer that migrate down from higher passes
with the first snowfall. On southeastern Sierran slopes they have proven
to be a preferred food of Tule Elk (*Cervus elaphus nannodes*).

In earlier days, Antelope Bush was a staple of vast Pronghorn herds
that roamed nomadically about California. The rapid development of
the state and uncontrolled shooting of wildlife reduced and restricted
Pronghorn herds to the sage lands of its northeastern corner, where they
largely remain today. Since 1982, the California Department of Fish
and Game has been reintroducing these fastest North American mammals onto former historic ranges, where they are increasing in numbers.

In the Pronghorn's day, grass was a major component of eastern

California's sagebrush lands, offering a diversified food choice for wildlife. Overgrazing by cattle and domestic sheep wiped out many of the grasses, although wild Rye Grass (*Elymus* spp.) and others survive in places.

Two shrubs add seasonal color to the otherwise gray-green tone of the Pinyon–Juniper Woodland shrub understory, Desert Peach (*Prunus andersonii*) and Rubber Rabbitbrush. Desert Peach literally blankets its bushy 3 to 6-foot high form (1 to 2 m) with striking deep rose flowers for two weeks in late April. Some of these turn into small, fuzzy, inedible "peaches." The shrub usually spreads by clones, connecting through underground stems.

The really brilliant color comes from rabbitbrush whose round clumps, 3 to 6 feet high (1 to 2 m), are covered with bright golden flower heads in late summer and autumn. The many extremely variable species and subspecies of the shrub grow over much of the West in all types of habitats, from alkali flats to sagebrush and Pinyon–Juniper Woodland and Mixed Conifer Forests.

A pioneer in invading disturbed areas, rabbitbrush, unlike Big Sagebrush and Antelope Bush, resprouts readily after a fire and hence takes over some of their sites. Its seeds and silvery green leaves are eaten by birds, rabbits, and small mammals, and its leaves are a winter food source for deer.

Conserving Water

The understory shrubs of the Pinyon–Juniper Woodland share essential survival traits with the trees in arid Great Basin country. Among their ways of conserving scarce and precious water, some carry small, waxy leaves. Rabbitbrush has leaves, branchlets, and stems all covered with whitish, feltlike fine hairs that form an insulating air trap and reduce evaporation. Big Sagebrush and Antelope Bush leaves show similar hair layers. In addition, the whiteness of the hairs reflects light that would otherwise increase the heat load and consequent water loss. Shrubs like the ephedras are leafless, their yellowish green stems carrying on foodmaking with minimal water loss. The price paid for all this is a reduced growth rate: no "free lunch" exists in nature.

The water-conserving adaptations of Pinyon–Juniper Woodland plants also help fulfill the water needs of the wildlife. Some plant foods

Figure 36. Ord's Kangaroo Rat lives in sagebrush flats that
intermingle with Pinyon Pine–Juniper Woodland. It
emerges from its burrows at night to feed.

are highly fluid. Fruits may be 90 percent water, and buds, leaves, new
twigs, and flowers almost as moist, whereas dry seeds contain a bare 5
to 10 percent.

Insects that eat each other, or plants, provide a nice wet food for
their predators, 50 to 90 percent liquid. And the Pinyon Mouse, jackrabbit, or ground squirrel that a Red-tailed Hawk (*Buteo jamaicensis*)
zeros in on for dinner is generally around 67 percent juice.

Ord's Kangaroo Rat (*Dipodomys ordii*) solves its water needs in several unique ways. The long-tailed, nocturnal rodent spends the hot,
drying days of summer in its moist burrow and comes out at night into
the cool air. As it hops about on its long hind legs, the kangaroo rat's
large, luminous eyes search for favorite dry seeds. Its body can convert
these dry seeds internally into metabolic water and produce a highly concentrated urine with minimal water loss. Brewer's and Black-throated
Sparrows (*Amphispiza bilineata*) are among the birds which survive in
this same way.

The kangaroo rat possesses an additional recycling mechanism. As it sleeps in its tightly sealed burrow during the day, the moisture from its own breath is trapped inside and absorbed by the seeds stored there.

The streams that flow down the eastern Sierra Nevada slopes past clumps of willows, Black Poplars (*Populus trichocarpa*), and Water Birch (*Betula occidentalis*), on through the pinyon-juniper belt, provide drinking water for the creatures that live near them or can travel to them. But the majority, far removed from such riparian haunts, must have their own solutions or perish. The wildlife of the Pinyon Pine–Juniper Woodlands and intermingled sage flats, from Sage Grouse to Pinyon Jays to kangaroo rats, show the adaptations necessary for successful existence in their land of little rain.

12

The Klamath Region

One of the most remarkable stories in the evolution of the state's forests has unfolded in the Klamath region of northwestern California and southwestern Oregon.

The Klamath region, a geologically complex, mountainous expanse of rugged peaks, deeply dissected by narrow valleys and wild rivers, contains rocks older than those of the Coast Ranges and Cascades. Its mountains rise to 9,000-foot heights (2,700 m), hold an incredibly rich flora, including conifers found nowhere else in the state, and house an exceptional variety of habitats. The diversity of the mountains, their varying exposures, soils, and local climates, combine to make this area one of the most exciting botanical frontiers in California.

Part of the reason for the unusual diversity lies in the misty past. From fossil diggings in Weaverville and Santa Rosa, California, and Oregon and Nevada, Daniel Axelrod has pieced together a sequence of major landscape changes in the West over long spans of time. Some of these bear directly on the Klamath region.

Thirty million years ago in the Tertiary period, much of the West was a region of rolling plains and low mountains. No major mountain chains existed, and the climate overall was mild to moderate, with ample summer rains. In this favorable setting an extremely rich mixture of conifers and deciduous hardwood trees thrived. Native western United States species such as spruce, pine, fir, larch, cedar, Douglas-Fir, Giant Sequoia, Redwood, Madrone, Tan Oak, alder, maple, poplar, oak, and aspen intermingled with native eastern U.S. species such as Sassafras,

Map 12. The Klamath region.

Beech, Sweet Gum, magnolia, chestnut, hawthorn, and locust, and with trees today native only to China such as Ginkgo, Dawn Redwood, and Tree-of-Heaven. These moisture-loving forests were the most diversified forests ever recorded for the temperate zone.

The climate began to change around 15 million years ago. The regular summer rains gradually disappeared and the winter rains diminished. The land became dryer. Summer temperatures rose. Trees that required the former wetness died off. Ginkgo, Dawn Redwood, and Tree-of-Heaven vanished. Many of the hardwoods remained only in the eastern United States, which still supplied the summer rains they needed. The conifers that survived were reduced to pockets on the cooler, higher slopes, with many spruce, fir, larch, and cedar confined farther north.

When the Sierra Nevada and Cascade ranges pushed upward during the last 10 million years, they brought other major changes. Their 14,000-foot heights (4,200 m) captured Pacific Ocean moisture on their western slopes and cast a gigantic rain shadow to the east. This created desert conditions that effectively eliminated most forest remnants.

The newly opened dry areas east of the mountains welcomed shrubs, grasses, and arid-adapted trees that migrated via windblown and animal-carried seeds from the Southwest. Giant Sequoias became fossils in Nevada but survived in moist, cool groves on the western midmountain slopes of the Sierra Nevada.

The remaining conifers from both the earlier forests and from later Pleistocene forests (2 million years ago) found suitable habitats principally in the middle and higher mountains. The richest collection of them ended up in the only section of California that retained a climate resembling that of Tertiary times—the Klamath region.

Today, the Klamath region serves both as a refugium for conifers and other plants from earlier eons and as an arena for active diversification of current plant life. This combination results in a rich mixture of species.

The Klamath River, home of legendary steelhead and salmon, cuts a gorge through the heart of the region, separating the Siskiyou Mountains on the north and west from the Marble and Salmon Mountains in the center. The Salmon Mountains merge with the Scott Mountains and Trinity Alps to the south.

The name Klamath Mountains covers them all and defies any general description. The western subregion lies close enough to the Pacific Ocean to feel maritime influence, receiving up to 125 inches (313 cm) of precipitation annually. The eastern interior subregion is drier. But within each subregion so many local pockets of divergent rock base, soil, soil moisture, elevation, and topography exist that the variety of plant life is continually surprising. In places, the accidents of history have left relicts. Elsewhere new plants have evolved and others have invaded from surrounding or distant territories. Sources and arrival times are not always clear.

Eastern Subregion

The southeastern and central Klamath forests look a lot like the forests of the Sierra Nevada. But a climb in certain parts of this region, such as the Russian Peak–Little Duck Lake section of the Salmon Mountains, produces some very un-Sierran discoveries.

The trail leads into superb lower elevation Mixed Conifer Forests of Ponderosa Pine, Sugar Pine, White Fir, Incense Cedar, and Douglas-Fir, similar to Sierran, growing on deep granitic soils. In the shady, moist forests Snow Plants send up scarlet shoots near the nodding bells of Pipsissewa, the tall, sticky stalks of Pine Drops, and the trailing leaves of delicate Twin-flowers (*Linnaea borealis*).

Further, on the same trail, increased elevation reveals a glimpse of the region's unique conifer diversity. At one point along Horse Range Creek, eleven different kinds of conifers are visible. Typical Sierran midmountain conebearers—Ponderosa Pine, Sugar Pine, White Fir, Douglas-Fir—intermingle with typical higher elevation conifers—Lodgepole Pine, Western White Pine, Mountain Hemlock, Shasta Red Fir. These are joined on moist, northeast-facing ravines by Pacific Yew; Engelmann Spruce (*Picea engelmannii*), a tree of the Rocky Mountains and Cascades; and Brewer Spruce (*Picea breweriana*), an exceptionally striking endemic found only in the Klamath region.

A continued climb past Chaparral thickets of Huckleberry Oak, Greenleaf Manzanita, and shiny-leaved Tobacco Bush (*Ceanothus velutinus*) leads to higher elevation forests dominated by Mountain Hemlock and Shasta Red Fir. Moist meadows with showy White-flowered Bog-orchids, the round white flower heads of Western Bistort (*Polygonum bistortoides*), and low-growing subalpine gentians (*Gentiana newberryi*) dot the high country.

In the meadow west of Little Duck Lake stand the tall narrow spires of Subalpine Fir (*Abies lasiocarpa*), a second Rocky Mountain species, discovered on Russian Peak by John Sawyer and Dale Thornburgh in 1968.

Halfway up a steep slope near timberline, trees with a very different shape appear. These are the rugged, thickly needled Foxtail Pines, found at treeline in California only on localized high Klamath peaks and in the high southern Sierra Nevada. Today's rare Foxtails are probably relict populations from a time when they grew continuously between the present sites.

The final ascent to the summit of Russian Peak, 8,196 feet (2,459 m), traverses a timberline woodland of Whitebark Pine and matlike Common Juniper (*Juniperus communis*).

Seventeen different kinds of conifers live within one square mile in the Russian Peak-Little Duck Lake area, the most concentrated collection of conifers on the planet. In addition, more than 450 other species of plants occur here, many found nowhere else.

Not all of the Klamath Mountains' riches climb steep slopes. West of Mount Shasta City, the trailhead to Cedar Lake offers hiking access to the Cedar Basin. There, on relatively level ground, thirteen kinds of conifers grow side by side. Most of these would be restricted to specific altitudinal zones elsewhere in the state. All but one are readily accessible to average hikers.

The inland form of Port Orford Cedar thrives there in a mixed forest of Lodgepole Pine, Jeffrey Pine, Sugar Pine, Ponderosa Pine, Western White Pine, Whitebark Pine, Mountain Hemlock, Incense Cedar, Douglas-Fir, White Fir, Brewer Spruce (requires a climb), and Shasta Red Fir.

Shasta Red Fir (*Abies magnifica* var. *shastensis*) is the primarily northern California variety of Red Fir, named for Mount Shasta, where it occupies upper forested levels. It differs from Red Fir chiefly by bracts that partly protrude from between its cone scales. Red Firs of the central Sierra have hidden bracts in their cones; those of Sequoia/Kings Canyon National Parks, however, also show the protruding bracts.

In the Klamath region Shasta Red Fir intergrades with Noble Fir (*Abies procera*), which continues north in the moist middle and high elevation forests of Washington and Oregon. Noble Fir carries the protruding cone bracts to extremes, extending prominent papery ones out and abruptly down from between the scales of each barrel-shaped cone.

Some Special Species

Brewer Spruce, often called Weeping Spruce, ranks among the special treasures of the Klamath region. Once more widespread over the West in Tertiary times, it now exists as a relict in the most hard-to-reach places. The tree was named for William Brewer, who passed through the Siskiyous in 1863 on his four-year Whitney Survey of California's natural resources.

Many Brewer Spruce prefer the slopes of steep, rocky, north-facing canyons at high Klamath altitudes, 6,000 to 8,000 feet (1,800 to 2,400 m), where snows supply water well into summer and where they are least susceptible to fire. They do, however, tolerate a variety of soil types and exposures in the Siskiyous and often invade Red Fir Forests. Although primarily higher elevation conifers, in the western subregion they are a part of lower elevation Douglas Fir–White Fir Forests. Near

Figure 37. Brewer Spruce, with distinctive outward curving branches, grows only in the Klamath region.

Happy Camp they dominate mixed forests of White Fir, Noble Fir, and Deer Oak (*Quercus sadleriana*).

The Brewer Spruce forests that are protected lie primarily in wilderness areas at upper elevations. Low and midelevation sites are chiefly unprotected and vulnerable to logging.

Brewer Spruce has been called by many the most beautiful spruce in the world. Willis Linn Jepson considered the appearance of the trees "so singularly different from that of any other conifer that they cannot ever be mistaken" (Jepson 1923:87). Rising gracefully to a spirelike crown up to 160 feet high (48 m), Brewer Spruce is usually clothed to the ground with branches. The lower branches curve horizontally down and outward, turning up at the tip. These horizontal branches carry dense streamers of needle-covered branchlets which hang vertically, like cords, creating an arresting drapery effect. The weeping aspect and soft rounded needles separate the Brewer Spruce readily from the other rare spruce of the Klamath region, Engelmann Spruce.

Engelmann Spruce has a tall, slim profile with stiff lateral branches and short, sharp needles that are four sided and roll easily between the fingers. As on all spruce, the needles leave behind small wooden pegs when they fall, giving old twigs a bumpy look. Engelmann Spruce grows mainly on terraces near streams and on other moist lower slopes of the Shasta Red Fir Forests, where it frequently associates with another conifer that is scarce in California, Subalpine Fir.

Subalpine Fir sends up a slim, sharply pointed crown in both forests and meadows and also sprawls as shrubby krummholz at timberline. An aggressive competitor in high shade, it takes over new ground by layering, developing roots and stems where its branches touch the ground, often under snow.

Unlike Brewer Spruce, which survives only in the Klamath region, Engelmann Spruce and Subalpine Fir are common elsewhere. Both grow widely distributed as major trees in the Rocky Mountains and Cascades. Recent evidence suggests that Subalpine Fir may have been introduced into the Klamath region during the past few thousand years by birds carrying seeds. If this is true for the fir, it would be equally possible for Engelmann Spruce and other species.

Western Subregion

The western subregion of the Klamath Mountains contributes heavily to the outstanding diversity of the entire region.

The moist, cool climate of the western Siskiyou, Marble, Salmon, and Trinity Alps mountain ranges produces some extremely lush forests. As in all of the Klamath region, soil and the underlying rock base primarily determine the vegetation pattern. Granitic and sedimentary soils generally support rich forests. Serpentine's deficient soil carries a sparse cover. Elevation, soil moisture, and topography bear important secondary influences.

While the Klamath Mountains do not show the distinct altitudinal belts of trees that occur in the Sierra Nevada and Cascades, they reveal a mild zonation in many areas where certain species tend to occupy particular levels.

The lower elevation western forests, adjacent to the mixed evergreen stands of the Coast Ranges, are dominated by impressive Douglas-Fir, some 200 feet high (60 m) and 8 feet wide (2 m). Madrone and Tan Oak fill in below the canopy, along with Mountain Dogwood, Big-leaf and Vine Maple. Sugar Pine, Ponderosa Pine, and Incense Cedar appear scattered throughout the forest. Port Orford Cedar and Pacific Yew follow the streams. Ample rain produces a verdant forest floor cover including Twin Flower, Vanilla Leaf, and Siskiyou Inside-out-Flower (*Vancouveria chrysantha*).

Higher up the western slopes, these lower forests grade into mid-elevation forests of White Fir and Douglas-Fir. Most of the mixed

Figure 38. Port Orford Cedar often follows the streams or grows near lake edge in the Klamath region.

evergreens remain as minor understory components, along with Giant Chinquapin and Canyon Live Oaks. Sugar Pine and Ponderosa Pine are relegated to the open drier spaces amid the increased shade of maturing forests.

As the White Fir midelevation forests gradually climb the slopes to 4,500 feet or higher (1,350 m), they merge into the next higher elevation forests dominated by Noble or Shasta Red Fir. The transition area between the two often holds the best of both communities—exceptionally diverse forests.

The Noble or Shasta Red Fir Forests form mature, closed-canopied stands on well-developed soils, sharing space with scattered Lodgepole Pine, Western White Pine, and Mountain Hemlock. All of these conifers merge almost imperceptibly into the next higher Mountain Hemlock zone. Within both of these higher elevation forests, the dominant tree takes over the major expanse. But a mosaic of habitats exists on south-facing slopes, rocky ridges, talus, moraines, and glaciated slopes with differing pockets of soil, supporting a wide variety of species.

Among the unusual trees and probable relicts of the fir and hemlock forests are two conifers whose major distribution lies farther north, Pacific Silver Fir and Alaska Cedar (*Cupressus nootkatensis*).

Pacific Silver Fir thrives in a few scattered localities of the Siskiyou and Marble Mountains, most often on steep, cool, moist, north-facing slopes between 5,000 and 6,200 feet (1,500 to 1,860 m). There it frequently grows in pure stands, again in mixed forests with Noble Fir,

Mountain Hemlock, and Mountain Alder (*Alnus incana* ssp. *tenuifolia*). On rocky, warmer exposures, Pacific Silver Fir joins White Fir, Douglas-Fir, Sugar Pine, Jeffrey Pine, Western White Pine, the occasional Incense Cedar, and Brewer Spruce in a virtual botanical museum.

Pacific Silver Fir well deserves its scientific name, *Abies amabilis*, meaning "lovely fir." A handsome tree up to 200 feet tall (60 m), with a slender spirelike tip, it sends out drooping branches covered with luxuriant needles that are shiny dark green above and silvery below.

Like other true firs, the Pacific Silver Fir produces a second type of foliage on the cone-bearing branches at the top of the tree. These upper needles are sharp, stout, and curved. The velvety purple cones that sit erect among them in cone years fall apart when mature but leave a spikelike axis standing aloft for another year or more.

The other relict conifer of the Klamath's western subregion, Alaska Cedar, is chiefly at home in coastal Alaska and British Columbia, and in high, cool, moist elevations of the Washington and Oregon Cascades. It reaches its lower limit in the Siskiyou Mountains.

The durability of the tree's wood is legendary. The Kwakiutl Indians of Vancouver Island have long carved it into haunting ceremonial masks. Paradise Inn, in the heart of Mt. Rainier National Park, was furnished early in the park's history with beautiful, fine-grained lumber from fire-killed Alaska Cedar snags which had weathered in that rainy climate for around 50 years.

In its central range, as a medium-sized forest tree on Mount Rainier, Alaska Cedar is instantly recognizable by its wilted appearance. The whiplike tip bends over, creating a limp silhouette. As the prickly, scale-like leaves age, they give the tree a yellowish tinge. Death seems imminent. But not so: this look is normal for Alaska Cedar. The tree often lives 1,000 years or more in favorable cool climate with summer rain.

In the Siskiyous, Alaska Cedar forms isolated stands at subalpine levels. As a shrub or tree, it survives in a variety of open habitats, including ridges, boulder-strewn slopes, wet meadows, and lake margins. Alaska Cedar spreads mainly by layering. Its western fossils date back millions of years.

Another holdover from earlier times is one of the world's rarest oaks. Shrubby Deer Oak, sometimes called Sadler's Oak, locally common in both subregions of the Klamath Mountains, occurs nowhere else. With leaves more like a chestnut than an oak, and its nearest relatives in Japan, China, and the eastern U.S., this unusual plant fits the pattern of a remnant from the long ago luxuriant Tertiary forests.

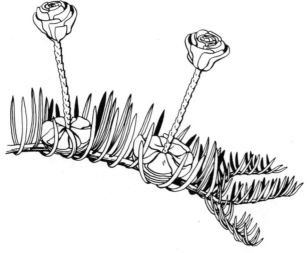

Figure 39. Fir cones disintegrate on the tree, shedding their
cone scales and seeds a few at a time.

Growing frequently in hedges, as Salal does along the coast, it often
reaches 7-foot heights (2 m). Its acorns rank high in the fare of chipmunks, squirrels, bears, deer, Band-tailed Pigeons, and other wildlife.
At Cook and Green Pass in the Siskiyous, Deer Oak thrives amid one
of the richest floras in the state. Copious lilies, wintergreens, heaths,
orchids, fungi, and ferns join the oaks under the conifers.

At Lake Eleanor in the Salmon-Trinity Alps, Deer Oak forms thickets in the next higher zone above the fragrant Western Azaleas and the
willows that line the lake shore. The oak's associates here include White
Fir, Western White Pine, Sugar Pine, Incense Cedar, Ponderosa Pine,
and Mountain Hemlock, a mixture unusual anywhere except in the
Klamath region.

A summer day on the shore of sylvan Lake Eleanor brings a touch
of Klamath serenity. The sound of the wind in the pines is interrupted
only by the flutelike downward spiral of the Hermit Thrush, the spir-
ited "Hic, three beers" of the Olive-sided Flycatcher, the loose trill of
Dark-eyed Juncos, and the clear, bright, notes of Fox Sparrows.

The lake itself holds active life. Nearly every Klamath lake hides amphibians such as the Rough-skinned Newts (*Taricha granulosa*), 6-inch
(15 cm) brownish salamanders with orange underparts. Crawling about
on the lake bottom, capturing insects and attaching eggs to horsetail
shoots, they operate with ease in water too cold for warm-blooded crea-

tures. Air breathers, they rise to the surface at intervals to take a gulp and return to the depths.

On rainy nights in fall, the newts migrate en masse back to land, doggedly climbing through tangles and litter to find wintering homes under logs, bark, or rotten wood. In spring they repeat the migration, moving with expressionless eyes and tireless undulations back to the same lake in which they hatched.

Red-legged Frogs (*Rana aurora*) and Western Toads (*Bufo boreas*) share many of these lakes. David Rains Wallace came across "a huge glistening mass of tiny, pot-bellied toads shuffling nose-to-tail" out of Little Elk Lake in the Marble Mountains on a frosty last day of summer. Thousands more of the black toad tadpoles swarmed in the warmer shallows, trying to become toads in a hurry, before the lake turned to ice (Wallace 1983:44).

Much of the same wildlife that inhabits California's North Coastal, Redwood, and Douglas-Fir/Mixed-Evergreen Forests lives in similar spots in the Klamath Mountains. The unique little Tailed Frog finds a home in small, frigid streams of higher areas. Dippers, Otters, Western Pond Turtles (*Clemmys marmorata*), and yellow-legged frogs frequent the rivers. Bald Eagles and Peregrine Falcons hunt by day, Northern Spotted Owls at night. Black Bears, Mountain Lions, Martens, and Fishers prowl the forests and canyons.

Pristine, inaccessible sections of Klamath wilderness may well be among the last bastions of the rarest mammal in California, the Wolverine (*Gulo gulo*).

Seeps

Throughout the Klamath region, where perennial running water overlies serpentine strata, seeps fill with the curious forms of the California Pitcher Plant (*Darlingtonia californica*). The brown mummified remains of last year's plants pack together tightly with fresh green ones in a solid mass, rising at different heights from 1 to 2 feet (60 cm) out of the soggy muck.

Found only in northern California and southern Oregon, this insectivorous plant has a long history in the area. Another Tertiary relict, it favors nutrient-deficient soils.

The plant's green leaves form hooded, elongated tubes, narrow at the bottom where they enclose a pool of liquid. The expanded hood at the top has given rise to the graphic name Cobra Lily. Insects crawl inside the hood to explore. There they find downward-pointing hairs and slippery surfaces. If they follow the hairs and then decide to turn around, they face a wall of bristles. They tire and eventually fall into the water, where enzymes digest their soft parts.

Insect-eating plants are rarities in the world, and this one, like others in the eastern United States and Europe, utilizes the nutrients it absorbs from the insects to make up for the deficiencies, especially nitrogen, in its soils. But the pitcher plant does not depend totally on trapped insects for survival. It photosynthesizes its own food and sends up separate stalks of flowers with creamy sepals and maroon petals.

Its Klamath seeps include other unique plants. Thickets of Western Azaleas, Labrador Tea, and Bog Laurel (*Kalmia microphylla*), with tiny magenta flowers, often crowd the lilies, grasses, and Stream Orchids (*Epipactus gigantea*) scattered among the carnivorous plants. On a miniature scale, the tiny rosettes of the Sundew (*Drosera rotundifolia*) capture their share of insects in the sticky hairs of their leaves.

Serpentine

Serpentine is widespread in California, but the most extensive outcroppings of the rock occur in the Klamath Mountains. Serpentine (peridotite) breaks down into soil that is impoverished in calcium, nitrogen, sodium, and potassium and unusually high in magnesium and the heavy metals of nickel, cobalt, and chromium. This potentially toxic mixture devastates most plants.

Yet, as always, in the natural competition for places to live, conditions that close the door for some species open it for others. Among the low-growing wild flowers that are Klamath serpentine indicators are the Serpentine Arnica (*Arnica cernua*), a yellow-flowered member of the sunflower family; the scarlet Short-lobed Indian Paintbrush (*Castilleja hispida*); and Howell's Lomatium (*Lomatium howellii*), a flatheaded yellow flower of the carrot family. All these have relatives that live elsewhere on "normal" soils.

When serpentine crops out in coniferous forests of the Klamath Mountains, its effect is usually dramatic. The surrounding mature, fully developed conifers are abruptly replaced by sparse, open stands of compact or stunted Jeffrey Pines and Incense Cedar. The herbs and shrubs are usually dwarfed. Leaves are smaller and tougher, with a waxy or hairy bloom to reduce water loss from the hot, reflective soil surface. One common indicator, Leather Oak (*Quercus durata*), evergreen and up to 9 feet high, covers its stiff twigs with a dense, white wooly mat.

Where more moisture is available, Port Orford Cedar and Western White Pine grow on serpentine in ravines or near water. They are replaced on the dry ridgetops by Chaparral species like Tan Oak, silk tassel bush (Garrya spp.), Huckleberry Oak, and shrubby California Bay.

Coexisting with the plants are insects equally adapted to surviving the high summer surface temperatures and to feeding on plant leaves containing extraordinary levels of heavy metals.

Although serpentine rock is lustrous green, it weathers to pale red and turns some of the Klamath region into "red rock country."

The richness of the forests in the Klamath region results from a combination of many factors. Geographically the Klamath Mountains form a loose "bridge" between the Cascade—Sierra Nevada chain to the east and the Coast Ranges to the west. Ecologically they carry their own distinctive mixture of flora from both sources, plus a northern element, plus relicts from the Tertiary and Pleistocene, plus bird-carried introductions.

The great diversity of topography, from rugged peaks to narrow valleys, steep slopes, and all exposures of the compass, insures endless varieties of habitats. The differing rock types and soils, seeps, and serpentine domes, drier areas to the east, moister areas to the west, all are harbingers of a multiplicity of life. And the climate, with a humidity greater than the northern Sierra, a precipitation starting earlier and lasting longer, with a little summer rain over the higher slopes—all this creates conditions more nearly like those of rich Tertiary times than in any other montane area in California.

13

Battle Lines

California's onetime unmatched pristine forests have lost devastating amounts of their wildness to logging, clear-cutting, herbicides, overuse and development in the past 150 years. Still the finest of their kind in the temperate zone, the state's forests continue to be a battleground for opposing interests.

In the early 1990s, as more than one-fourth of California's plants and animals became endangered species and one-half of the state's habitats were imperiled by human impact, concerned state and federal agencies joined forces with citizens groups on a plan to preserve California's biodiversity before it is too late. The new emphasis focuses on preserving the diversity of bioregions, natural areas large enough to include watersheds and all pertinent geographic features for flora and fauna, rather than the fragmented areas of the past.

Ecologically minded forest managers have been aware for years of the need to maintain the biodiversity of their forests—to encourage the full spectrum of plant and animal life native to each forest. The natural places we consider beautiful are nearly all diversified, functioning ecosystems, whether they be Giant Sequoia groves, Red Fir Forests, mountain meadows, or insectivorous plant seeps.

The dynamic balance of a healthy ecosystem both produces superb natural areas and builds a reservoir of the genetic potential of all their species. Evolution uses such varied gene pools, over time, to select survivors by trial and error. We use them for pragmatic purposes, frequently medical.

189

Map 13. California counties. From *California Mammals* by E. W. Jameson, Jr., and Hans J. Peeters, illustrations by Hans J. Peeters; University of California Press, 1988.

More than 70 percent of pharmaceutical drugs available today come from plants. Salicin, the bitter-tasting substance from which aspirin is derived, came originally from the leaves of the myrtle tree and the bark of a willow. With only about 1 percent of the world's trees screened for medicines, the potential for important discoveries is enormous.

The finding of taxol in California forests in 1991 was one such bonanza. In a random sampling of western trees, scientists from the National Cancer Institute learned that taxol from the bark and needles of the Pacific Yew was effective in treating ovarian cancer. Until this discovery, yew had been considered a scraggly nuisance by northern California loggers, used as a cushion on which to drop big trees. Future cures for other kinds of cancer, cardiovascular diseases, the common cold, or AIDS may lie undetected in forests yet to be explored.

Recently, Chimpanzees in Tanzania were observed chewing the pith of certain trees never a part of their normal diet. They spit out all but the bitter, foul-smelling, bright red oily juice. Chemical analysis showed the juice to contain an ingredient that killed both fungal infections and many kinds of parasitic worms. Among the vulnerable parasites were the causes of schistosomiasis, a disease which devastates millions of people in the tropics.

The medicinal potentials make a strong argument for preserving forests and all the diversified plants and animals that have evolved with them over the ages. The health of the trees themselves often depends on the presence in their gene pool of pest-resistant strains and on genes giving the ability to adapt to environmental changes.

Pines are proving to be much more sophisticated genetic entities than was once supposed. Trees replanted in mountains after fires do much better if they come from the same areas and elevations as the originals. When the Institute of Forest Genetics compared Sugar Pines from various parts of California, it found that the trees from Junipero Serra Peak possessed an unusually high resistance to blister rust, whereas the trees from Sierra San Pedro Martir in Baja California had unusually blue foliage, often useful in fighting drought. When you've seen one Sugar Pine, you haven't seen them all!

With all the variations and unknowns in forest genetics and ecology, Aldo Leopold's staunch precept, "to keep every cog and wheel is the first precaution of intelligent tinkering," makes ever more sense. There is much yet to learn from forests where everything forms a piece of the whole. In a world of always varying ideas on what is the right use of the land, Leopold was decades ahead of his time: "a thing is right when it tends to preserve the integrity, stability and beauty of the biotic community" (Leopold 1966: 190, 262).

The battles to save forests, in California, in the tropics, and worldwide, go on and on. Only the battle lines change from year to year. Many of the big issues are being fought out in the courtrooms. Public awareness of what is really happening comes chiefly through support of environmental organizations that are in the thick of the fray. For those who wish to come aboard, I list below some of the prime defenders of California forests.

California Native Plant Society
1722 J Street, Suite 17
Sacramento, California 95814

California Nature Conservancy
785 Market Street
San Francisco, California 94103

California Wilderness Coalition
2655 Portage Bay East, Suite 5
Davis, California 95616

National Audubon Society
555 Audubon Place
Sacramento, California 95825

Planning and Conservation League
926 J Street, Suite 612
Sacramento, California 95814

Sierra Club
P.O. Box 7959
San Francisco, California 94120-9943

Sierra Club Legal Defense Fund
180 Montgomery Street, Suite 1400
San Francisco, California 94104

The Wilderness Society
900 Seventeenth Street, N.W.
Washington, D.C. 20006-2596

Selected References

Amme, David. 1977. The Torrey Pine and Its Big Cone Relatives. *Fremontia* 4(4):10–14.

Arno, Stephen F. 1973. *Discovering Sierra Trees*. Three Rivers, Calif: Yosemite Natural History Association and Sequoia Natural History Association.

———. 1977. *Northwest Trees*. Seattle: The Mountaineers.

Axelrod, Daniel I. 1976. *History of the Coniferous Forests, California and Nevada*. Berkeley and Los Angeles: University of California Publications in Botany.

Bakker, Elna. 1984. *An Island Called California*. Berkeley and Los Angeles: University of California Press.

Barbour, Michael G., and Jack Major. 1988. *Terrestrial Vegetation of California*. Expanded edition. Sacramento: California Native Plant Society.

Barbour, Michael G., et al. 1993. *California's Changing Landscapes*. Sacramento: California Native Plant Society.

Barrows, Cameron. 1985. Cool Owls of the Old Forests. *Pacific Discovery*. April.

Barrows, Katherine. 1984. Old-Growth Douglas-Fir Forests. *Fremontia* 11(4): 20–23.

Basey, Harold E. 1976. *Discovering Sierra Reptiles and Amphibians*. Three Rivers, Calif.: Yosemite Natural History Association and Sequoia Natural History Association.

Becking, Rudolf W. 1982. *Pocket Flora of the Redwood Forest*. Covelo, Calif.: Island Press.

Beedy, Edward Crosby. 1981. "Bird Community Structure in Coniferous Forests of Yosemite National Park, Calif." Ph.D. diss., University of California, Davis.

Berrey, Henry. 1987. Burning the Sequoias. *Yosemite*. 49(3):1–5.

Bingham, Bruce B., et al. 1991. Distinctive Features and Definitions of Young, Mature, and Old-Growth Douglas-fir/Hardwood Forests. In *Wildlife and Vegetation of Unmanaged Douglas-fir Forests*. USDA Forest Service, Gen. Tech. Rept. PNW-GTR-285. Pacific NW Research Station, Portland.

Biswell, Harold H. 1989. *Prescribed Burning in California Wildlands Vegetation Management*. Berkeley and Los Angeles: University of California Press.

Bock, Carl E., and James F. Lynch. 1970. Breeding Bird Populations of Burned and Unburned Conifer Forest in the Sierra Nevada. *The Condor* 72(2): 182–189.

Bull, Evelyn L., and Mark G. Henjum. 1987. The Neighborly Great Gray Owl. *Natural History*. Sept.

Byers, John A. 1989. Pronghorns in and out of a Rut. *Natural History*. April.

Davies, John. 1980. *Douglas of the Forests, the North American Journals of David Douglas*. Seattle: University of Washington Press.

Davis, Douglas F., and Dale F. Holderman. 1980. *The White Redwoods: Ghosts of the Forest*. Happy Camp, Calif.: Naturegraph Publishers, Inc.

Denison, William C. 1973. Life in Tall Trees. *Scientific American*. June.

DeSante, Dave. 1985. What Controls the Numbers of Subalpine Birds? *Newsletter 71* (Winter): 1–5. Stinson Beach, Calif.: Point Reyes Bird Observatory.

Dunn, Anthony T. 1985. The Tecate Cypress. *Fremontia* 13(4): 3–7.

Engbeck, Joseph H., Jr. 1973. *The Enduring Giants*. Berkeley: University Extension, University of California.

Evens, Jules G. 1988. *The Natural History of the Point Reyes Peninsula*. Point Reyes, Calif.: Point Reyes National Seashore Association.

Finson, Bruce. 1977. Staircase through Time. *Pacific Discovery*. Nov. Jughandle State Reserve.

Franklin, Jerry. 1989. Toward a New Forestry. *American Forests*. Nov.

Franklin, Jerry F., et al. 1981. Ecological Characteristics of Old-Growth Douglas-Fir Forests. USDA Forest Service, Gen. Tech. Rept. PNW-118. Pacific NW Forest and Range Experiment Station, Portland.

Gaines, David. 1988. *Birds of Yosemite and the East Slope*. Lee Vining, Calif.: Artemesia Press.

Gilbert, Bil. 1979. A Somewhat Peculiar Fellow. *Audubon*. Sept.

Griffin, James R. 1972. What's So Special About Huckleberry Hill on the Monterey Peninsula? In *Forest Heritage, A Natural History of the Del Monte Forest*, compiled by Beatrice F. Howitt, 3–7. Sacramento: California Native Plant Society.

Griffin, James R. 1988. Oak Woodland. In *Terrestrial Vegetation of California*, expanded edition, ed. Michael G. Barbour and Jack Major, 383–415. Sacramento: California Native Plant Society.

Griffin, James R., and William B. Critchfield. 1973. *The Distribution of Forest Trees in California*. Research Paper PSW-82. Berkeley: USDA Forest Service, California.

Hargis, C. D. 1982. *Winter Habitat Utilization and Food Habits of Pine Martens in Yosemite National Park*. Technical Report No. 6, Cooperative National Park Resources Studies Unit, University of California, Davis.

Hartesveldt, Richard J., et al. 1975. *The Giant Sequoia of the Sierra Nevada*. Washington, D.C.: National Park Service, U.S. Dept. of the Interior.

Harvey, H. Thomas, Howard S. Shellhammer, and Ronald E. Stecker. 1980. *Giant Sequoia Ecology*. Washington D.C.: National Park Service, U.S. Dept. of the Interior.

Harvey, H. Thomas, et al. 1981. *Giant Sequoias.* Three Rivers, Calif.: Sequoia Natural History Association.

Hickman, James C., ed. 1993. *The Jepson Manual: Higher Plants of California.* Berkeley: Jepson Herbarium and Library, University of California.

Hopkins, Natalie. 1982. Mycorrhizae: Secret Partnership of Plants and Fungi. *Fremontia* 10(2):11–15.

Hubbs, Carl L., and Thomas W. Whitaker, eds. 1972. *Torrey Pines State Reserve.* La Jolla, Calif.: The Torrey Pines Association.

Hutchinson, Judy L., and G. Ledyard Stebbins. 1986. *A Flora of the Wright's Lake Area.* Pollack Pines, Calif.: Judy L. Hutchinson.

Ingles, Lloyd G. 1945. Nesting of the Goshawk in Sequoia National Park, California. *The Condor* 47(4):215.

Jameson, E. W., Jr., and Hans J. Peeters. 1988. *California Mammals.* California Natural History Guide No. 52. Berkeley and Los Angeles: University of California Press.

Jenny, Hans, R. J. Arkley, and A. M. Schultz. 1969. The Pygmy Forest-Podsol Ecosystem and Its Dune Associates of the Mendocino Coast. *Madroño* 20(2):60–74.

Jepson, Willis L. 1923. *The Trees of California.* Berkeley: Associated Students Store, University of California.

————. 1934. *Trees, Shrubs and Flowers of the Redwood Region.* San Francisco: Save the Redwoods League.

Johnston, Verna R. 1970. *Sierra Nevada.* Boston: Houghton Mifflin.

Keator, Glenn. 1986. A Grand Tour of Northern California's Conifers, Part I. *Fremontia* 14(1):19–26.

————. 1988. A Tour of Northern California's Conifers, Part II. *Fremontia* 15(4):10–16.

Kelly, John. 1987. Saving the Forest Primeval: Report on the Spotted Owl. *Wingbeat,* Summer, 9–10. Stinson Beach, Calif.: Point Reyes Bird Observatory.

Koford, Rolf R. 1982. Mating System of a Territorial Tree Squirrel (*Tamiasciurus douglasii*) in California. *Journal of Mammalogy* 63(2):274–283.

Kruckeberg, Art. 1984. California's Serpentine. Part I. *Fremontia* 11(4): 11–17.

————. 1984. The Flora on California's Serpentine, Part II. *Fremontia* 11(5): 3–10.

Lanner, Ronald M. 1981. *The Piñon Pine: A Natural and Cultural History.* Reno: University of Nevada Press.

————. 1984. *Trees of the Great Basin: A Natural History.* Reno: University of Nevada Press.

Latting, June, ed. 1976. *Plant Communities of Southern California.* Special Publication No. 2. Sacramento: California Native Plant Society.

Lederer, Roger J. 1977. Winter Territoriality and Foraging Behavior of the Townsend's Solitaire. *American Midland Naturalist* 97(1):101–109.

Ledig, F. Thomas. 1984. Gene Conservation, Endemics and California's Torrey Pine. *Fremontia* 11(3):9–13.

Leopold, Aldo. 1966. *A Sand County Almanac, With Essays on Conservation from Round River.* New York: Oxford University Press, Inc.

Ligon, J. David. 1978. Reproductive Interdependence of Piñon Jays and Piñon Pines. *Ecological Monographs* 48(2):111–126.

Lyons, Kathleen, and Mary Beth Cuneo-Lazaneo. 1988. *Plants of the Coast Redwood Region*. Los Altos, Calif.: Looking Press.

MacRoberts, Michael H. 1974. Acorns, Woodpeckers, Grubs and Scientists. *Pacific Discovery*. Oct.

Maser, Chris, and James M. Trappe, eds. 1984. *The Seen and Unseen World of the Fallen Tree*. Gen. Tech. Rept. PNW-164. Washington, D.C.: USDA Forest Service.

Maser, Chris, James M. Trappe, and Ronald A. Nussbaum. 1978. Fungal-Small Mammal Interrelationships with Emphasis on Oregon Coniferous Forests. *Ecology* 59(4):799–809.

Maser, Zane, Chris Maser, and James M. Trappe. 1985. Food Habits of the Northern Flying Squirrel (*Glaucomys sabrinus*) in Oregon. *Canadian Journal of Zoology* 63(5):1084–1088.

Medley, Steven P., ed. 1987. Great Gray Owl Research in High Gear. *Update*, Spring, 1–2. The Yosemite Association.

McGinnis, Bridget. 1991. Great, Gray and Mysterious. *Yosemite* 53(2):1–4. The Yosemite Association.

Mirov, Nicholas T., and Jean Hasbrouck. 1976. *The Story of Pines*. Bloomington: Indiana University Press.

Mozingo, Hugh N. 1987. *Shrubs of the Great Basin: A Natural History*. Reno: University of Nevada Press.

Muir, John. 1901. *Our National Parks*. Boston: Houghton Mifflin.

——. 1961. *The Mountains of California*. New York: Doubleday and Company, Inc.

Munz, Philip A. 1963. *California Mountain Wildflowers*. Berkeley and Los Angeles: University of California Press.

——. 1968. *Supplement to A California Flora*. Berkeley and Los Angeles: University of California Press.

Munz, Philip A., and David D. Keck. 1959. *A California Flora*. Berkeley and Los Angeles: University of California Press.

Murray, Marshall D. 1989. Conifer Forests in the San Gabriel Mountains. *Fremontia* 17(3):11–14.

Niehaus, Theodore F., and Charles L. Ripper. 1976. *A Field Guide to Pacific States Wildflowers*. Boston: Houghton Mifflin.

Ornduff, Robert. 1974. *Introduction to California Plant Life*. California Natural History Guide No. 35. Berkeley and Los Angeles: University of California Press.

Parmeter, John R., Jr. 1979. *Fomes annosus* in California Forests. *California Agriculture* 35(5):27–28.

Pavlik, Bruce M., et al. 1991. *Oaks of California*. Sacramento: Cachuma Press and the California Oak Foundation.

Peattie, Donald Culross. 1953. *A Natural History of Western Trees*. Boston: Houghton Mifflin.

Peterson, P. Victor. 1966. *Native Trees of Southern California*. California Natural History Guide No. 14. Berkeley and Los Angeles: University of California Press.

Peterson, P. Victor, and P. Victor Peterson, Jr. 1975. *Native Trees of the Sierra Nevada*. California Natural History Guide No. 36. Berkeley and Los Angeles: University of California Press.

Peterson, Roger T. 1990. *A Field Guide to Western Birds*. Boston: Houghton Mifflin.

Powell, Jerry A., and Charles L. Hogue. 1979. *California Insects*. California Natural History Guide No. 44. Berkeley and Los Angeles: University of California Press.

Raven, Peter H. 1988. The California Flora. In *Terrestrial Vegetation of California*, expanded edition, ed. Michael G. Barbour and Jack Major, 109–135. Sacramento: California Native Plant Society.

Roof, James. 1975. Floral Watch on Siskiyou County's Cook and Green Pass. *The Four Seasons*. 5(1):16–23. Berkeley: Regional Parks Botanic Garden.

Russell, Ward C. 1947. Mountain Chickadee Feeding Young Williamson's Sapsuckers. *The Condor* 49(2):83.

Ryker, Lee C. 1984. Acoustic and Chemical Signals in the Life Cycle of a Beetle. *Scientific American*. June.

Ryser, Fred A., Jr. 1985. *Birds of the Great Basin: A Natural History*. Reno: University of Nevada Press.

Rundel, Philip, David J. Parsons, and Donald T. Gordon. 1988. Montane and Subalpine Vegetation of the Sierra Nevada and Cascade Ranges. In *Territorial Vegetation of California*, expanded edition, ed. Michael G. Barbour and Jack Major, 559–599. Sacramento: California Native Plant Society.

Sawyer, J. O., and J. P. Smith, Jr. 1973. The Klamath Region. *Newsletter* 8(4):3–6. Sacramento: California Native Plant Society.

Sawyer, John O., Jr., and Dale A. Thornburgh. 1988. Montane and Subalpine Vegetation of the Klamath Mountains. In *Terrestrial Vegetation of California*, expanded edition, ed. Michael G. Barbour and Jack Major, 699–732. Sacramento: California Native Plant Society.

Sawyer, John O., Dale A. Thornburgh, and James R. Griffin. 1988. Mixed Evergreen Forest. In *Terrestrial Vegetation of California*, expanded edition, ed. Michael G. Barbour and Jack Major, 359–381. Sacramento: California Native Plant Society.

Schnell, Jay H. 1958. Nesting Behavior and Food Habits of Goshawks in the Sierra Nevada of California. *The Condor* 60(6):377–403.

Sequoia National History Association. 1988. Air Pollution Significant Threat. *Seedlings* 5(1):1–3. Newsletter of The Sequoia Natural History Association.

Sherman, Paul W., and Martin L. Morton. 1979. Four Months of the Ground Squirrel. *Natural History*. July.

Sholars, Robert. 1984. The Pygmy Forest of Mendocino. *Fremontia* 12(3):3–8.

Sigg, Jacob. 1983. The Foxtail Pine of the Sierra. *Fremontia* 11(1):3–8.

Smith, Christopher C. 1978. Structure and Function of the Vocalizations of Tree Squirrels (*Tamiasciurus*). *Journal of Mammalogy* 59(4):793–808.

———. 1981. The Indivisible Niche of *Tamiasciurus*: An Example of Nonpartitioning of resources. *Ecological Monographs* 51(3):343–363.

Spencer, Cathy. 1990. One Giant Leap. *Omni*. Oct.

Spencer, Wayne D., and William J. Zielinski. 1983. Predatory Behavior of Pine Martens. *Journal of Mammalogy* 64(4):715–717.

Stacey, Peter B., and Walter D. Koenig. 1984. Cooperative Breeding in the Acorn Woodpecker. *Scientific American.* Aug.

Stebbins, Robert C. 1985. *A Field Guide to Western Reptiles and Amphibians.* Boston: Houghton Mifflin.

Stocking, Stephen K., and Jack A. Rockwell. 1989. *Wildflowers of Sequoia and Kings Canyon National Parks.* Three Rivers, Calif.: Sequoia Natural History Association.

Stone, Edward C., and Richard B. Vasey. 1968. Preservation of Coast Redwood on Alluvial Flats. *Science* 159(3811):157–161.

Storer, Tracy I., and Robert L. Usinger. 1964. *Sierra Nevada Natural History.* Berkeley and Los Angeles: University of California Press.

Sudworth, George B. 1967. *Forest Trees of the Pacific Slope.* New York: Dover Publications.

Sutherland, G. D., et al. 1982. Feeding Territoriality in Migrant Rufous Hummingbirds; Defense of Yellow-bellied Sapsucker (*Sphyrapicus varius*) Feeding Sites. *Canadian Journal of Zoology* 60(9):2046–2050.

Thomas, John Hunter, and Dennis R. Parnell. 1974. *Native Shrubs of the Sierra Nevada.* California Natural History Guide No. 34. Berkeley and Los Angeles: University of California Press.

Thorne, Robert F. 1988. Montane and Subalpine Forests of the Transverse and Peninsular Ranges. In *Terrestrial Vegetation of California,* expanded edition, ed. Michael G. Barbour and Jack Major, 537–557. Sacramento: California Native Plant Society.

Todd, Paul. 1989. In Search of the Elusive Sierra Mountain Beaver. *Yosemite* 51(3):1–3,10, 12.

Tomback, Diana F. 1982. Dispersal of Whitebark Pine Seeds by Clark's Nutcracker: A mutualism hypothesis. *Journal of Animal Ecology* 51(2):451–467.

Trappe, James M., and Robert D. Fogel. 1978. Ecosystematic Functions of Mycorrhizae. In *The Belowground Ecosystem: A Synthesis of Plant-associated Processes,* ed. J. K. Marshall, 205–214. Series No. 26, Colorado State University Range Science Dept.

Trappe, James M., and Chris Maser. 1977. Ectomycorrhizal Fungi: Interactions of Mushrooms and Truffles with Beasts and Trees. In *Mushrooms and Man, an Inter-disciplinary Approach to Mycology,* ed. Tony Walters, 165–179. Forest Service, U.S. Dept. of Agriculture.

Tweed, William. 1987. Born of Fire. *National Parks.* Jan.

Udall, James R. 1986. Finis Mitchell, Lord of the Winds. *Audubon.* July.

VanderWall, Stephen, and Russell P. Balda. 1983. Remembrance of Seeds Stashed. *Natural History.* Aug.

Vasek, Frank C., and Robert F. Thorne. 1988. Transmontane Coniferous Vegetation. In *Terrestrial Vegetation of California,* expanded edition, ed. Michael G. Barbour and Jack Major, 797–832. Sacramento: California Native Plant Society.

Vogl, Richard J., et al. 1988. The Closed-cone Pines and Cypress. In *Terrestrial Vegetation of California,* expanded edition, ed. Michael G. Barbour and Jack Major, 295–358. Sacramento: California Native Plant Society.

Wake, David. 1990. Quoted in "Scientists Link Animal Deaths to Acid Snow" by Elliot Diringer. *San Francisco Chronicle*, Feb. 19, A5.

Wallace, David Rains. 1983. *The Klamath Knot*. San Francisco: Sierra Club Books.

West, Lorne. 1986. The Demise of Yosemite Valley's Evergreens. *Yosemite* 48(4):2.

Wheat, Margaret M. 1967. *Survival Arts of the Primitive Paiutes*. Reno: University of Nevada Press.

Whitney, Stephen. 1979. *A Sierra Club Naturalist's Guide to The Sierra Nevada*. San Francisco: Sierra Club Books.

————. 1985. *Western Forests*. New York: Alfred A. Knopf.

Wolf, Carl Edward. 1967. What's Bugging the Bristlecone Fir? *The Four Seasons* 2(3):2–5. Berkeley: Regional Parks Botanic Garden.

Yates, Steve. 1986. A Pronghorn Needs Freedom to Feel at Home on the Range. *Smithsonian*. December.

Young, Dorothy King. 1989. *Redwood Empire Wildflowers*. Happy Camp, Calif.: Naturegraph Publishers, Inc.

Zedler, Paul H. 1986. Closed-cone Conifers of the Chaparral. *Fremontia* 14(3): 14–17.

Zielinski, William J., et al. 1983. Relationship between Food Habits and Activity Patterns of Pine Martens. *Journal of Mammalogy* 64(3):387–396.

Zinke, Paul J. 1988. The Redwood Forest and Associated North Coast Forests. In *Terrestrial Vegetation of California*, expanded edition, ed. Michael G. Barbour and Jack Major, 679–698. Sacramento: California Native Plant Society.

Index

Hawk
  Red-tailed, 88, 174
  Swainson's, 151
Hazelnut, 97
Heartwood, illustration of, 52
Heather
  Mountain, 145
  White, 145
Hemlock
  distinguishing from Spruce and
    Douglas-Fir, 9–11
  Mountain, 144, 179, 180, 183, 185
  Western, 10, 28, 31–33
Heracleum lanatum, 22
Hermit Thrush, 98
Hermit Warbler, 98
  illustration of, 126
Heron, Great Blue, 88
Heterobasidion annosum, 101
Heteromeles arbutifolia, 81
High desert, woodlands of, 162–175
Horned Lizard, San Diego, 69
Horsetails, Giant, 22
House Finch, 68
House Wren, 85
Howell's Lomatium, 187
Huckleberry
  California, 23
  Evergreen, 62
  Red, 28
Huckleberry Hill, 65
Huckleberry Manzanita, 61
Huckleberry Oak, 129, 188
  illustration of, 131
Humboldt County, floods in, 15
Hummingbird
  Allen's, 66
  Anna's, 66
  Rufous, 66
Hutton's Vireos, 87
Hyphae, of mushroom, 38
Hypogeous, 40

Icterus galbula, 87
Incense Cedar, 7, 92, 95–96, 179–180,
  184–185
  illustration of, 8
Indian Paintbrush, Short-lobed, 187
Indians
  harvest of pine nuts, 164
  use of Alaska Cedar, 184
  use of Buckeye, 83
  use of Gray Pine, 77–78

  use of Juniper Woodland plants, 168
  use of Lodgepole Pine, 128
  use of moth larvae as food, 169–170
  use of pinyon pines, 164
  use of Poison Oak, 84
  use of Red Cedar, 31–33
  use of Sugar Pine, 93–94
Ingles, Lloyd, 133
Insect-eating plant, 186–187
Interior Live Oak, 43, 80–81
Invertebrates, in treetops, 48

Jack Pine, 72
Jackrabbit
  Black-tailed, 88, 150, 170
  White-tailed, 136, 150
Jay
  Florida Scrub, 87
  Mexican, 87
  Pinyon, 165–167
  Steller's, 24, 98, 132, 133
Jeffrey Pine, 103–104, 129, 180, 184,
  188
  association with Washoe Pine, 104–
    105
  distinguishing from Ponderosa Pine,
    105
  illustration of, 131
  moths of, 169–170
Jenny, Hans, 64–65
Jepson, Willis Linn, 21, 24, 181
Jepson Manual, The, 71, 78, 104
Johnny-jump-up, 80
Johnson, Robert Underwood, 120
Jughandle State Reserve, 63
Junco hyemalis, 61
Juncos, Dark-eyed, 61
Juniper
  Bennett, 145
  California, 81, 167, 168
  Common, 179
  distinguishing from cypress and cedar,
    7–9
  Sierra, 131, 145, 168
  Utah, 167, 168–169
  Western, 9, 103, 167
Juniper Woodland, 167–169
  birds of, 168
  mammals of, 168
Juniperus californica, 81, 168
Juniperus communis, 179
Juniperus occidentalis, illustration of, 9
Juniperus occidentalis var. *australis*, 168

Designer:	U. C. Press Staff
Compositor:	G&S Typesetters, Inc.
Text:	10/13 Galliard
Display:	Galliard
Printer:	Thomson-Shore, Inc.
Binder:	Thomson-Shore, Inc.